T0134907

Cognitive Intelligence and Robotics

Series Editors

Amit Konar, Department of Electronics and Telecommunication Engineering, Jadavpur University, Kolkata, India
Witold Pedrycz, Department of Electrical and Computer Engineering, University of Alberta, Edmonton, AB, Canada

Cognitive Intelligence refers to the natural intelligence of humans and animals, it is considered that the brain performs intelligent activities. While establishing a hard boundary that distinguishes intelligent activities from others remains controversial, most common behaviors and activities of living organisms that cannot be fully synthesized using artificial means are regarded as intelligent. Thus the acts of sensing and perception, understanding the environment, and voluntary control of muscles, which can be performed by lower-level mammals, are indeed intelligent. Besides the above, advanced mammals can perform more sophisticated cognitive tasks, including logical reasoning, learning, recognition, and complex planning and coordination, none of which can yet be realized artificially to the level of a baby, and thus are regarded as cognitively intelligent.

This book series covers two important aspects of brain science. First, it attempts to uncover the mystery behind the biological basis of cognition, with a special emphasis on the decoding of stimulated brain signals or images. Topics in this area include the neural basis of sensory perception, motor control, sensory-motor coordination, and understanding the biological basis of higher-level cognition, including memory, learning, reasoning, and complex planning. The second objective of the series is to publish consolidated research on brain-inspired models of learning, perception, memory, and coordination, including results that can be realized on robots, enabling them to mimic the cognitive activities performed by living creatures. These brain-inspired models of machine intelligence complement the behavioral counterparts studied in traditional artificial intelligence.

The series publishes textbooks, monographs, and contributed volumes.

More information about this series at http://www.springer.com/series/15488

Abhijit Mahapatra · Shibendu Shekhar Roy ·
Dilip Kumar Pratihar

Multi-body Dynamic Modeling of Multi-legged Robots

 Springer

Abhijit Mahapatra
Advance Design and Analysis Group
Central Mechanical Engineering
Research Institute
Durgapur, West Bengal, India

Shibendu Shekhar Roy
Department of Mechanical Engineering
National Institute of Technology Durgapur
Durgapur, West Bengal, India

Dilip Kumar Pratihar
Department of Mechanical Engineering
Indian Institute of Technology Kharagpur
Kharagpur, West Bengal, India

ISSN 2520-1956   ISSN 2520-1964   (electronic)
Cognitive Intelligence and Robotics
ISBN 978-981-15-2955-9   ISBN 978-981-15-2953-5   (eBook)
https://doi.org/10.1007/978-981-15-2953-5

This Springer imprint is published by the registered company Springer Nature Singapore Pte Ltd.
The registered company address is: 152 Beach Road, #21-01/04 Gateway East, Singapore 189721, Singapore

Dedicated to Our Family Members

Preface

Over the past two decades, multi-body dynamics has emerged as one of the major tools in the design and analysis of complex mechanical system. However, in the field of robotics, its application is not so comprehensive. Further, multi-legged robotics is an emerging technology and a lot of research is yet to be done before a full proof robust system is developed to fulfill the human needs. Most of the available literature about multi-body dynamic modeling of multi-legged robots is in form of journal, conference proceedings, monographs, etc. However, only a handful of books are available online or published in this field. This book deals with constrained rigid multi-body dynamics. It presents a systematic approach to kinematics and coupled multi-body dynamic modeling along with the study of generating path and gait of a realistic six-legged robotic system locomoting in 3D Cartesian space on various terrains like sloping surface, staircase, and some user-defined rough terrains with straight-forward, crab, and turning motion capabilities. The book outlines the most effective way of handling the kinematic and dynamic constraints and discusses in detail the foot–ground interaction mechanics, energy efficiency, and stability (both dynamic and static) of a realistic six-legged robot. Further, the book deals with computer simulations carried out in MATLAB and subsequent validation using Virtual Prototyping (VP) tools and real experiment.

Durgapur, India
Durgapur, India
Kharagpur, India

Abhijit Mahapatra
Shibendu Shekhar Roy
Dilip Kumar Pratihar

Acknowledgements

This book would not have been possible without the help of few people around us who mattered the most and without whom this work would not have been possible. The authors express their heartfelt thanks to Indra Bahadur Biswakarma, Hemant Agarwal, Sandip Tiwari, Kondalarao Bhavanibhatla, and Kaustav Biswas for their kind help and cooperation.

The first author would like to thank all his colleagues and friends at CSIR-Central Mechanical Engineering Research Institute, Durgapur, India, for their encouragement, cooperation, and assistance rendered during the course of the work. The second author is also thankful to all present colleagues of Mechanical Engineering Department at NIT, Durgapur, India. The third author is grateful to his present colleagues of Mechanical Engineering Department at IIT, Kharagpur, India.

The authors would like to especially thanks to Dr. Manuel Ferre, Dr. P. M. Pathak, as well as others who have contributed directly or indirectly to the content of the book.

The authors also thank the unknown copy-editor and the staff of Springer Nature for helpful cooperation and thorough publication of the book.

On a personal note, the authors would like to express deepest gratitude to their family members for their invaluable love, affection, and support.

Last but not least, authors would like to thank everybody who directly and indirectly helped them for the successful completion of this work.

Durgapur, India   Abhijit Mahapatra
Durgapur, India   Shibendu Shekhar Roy
Kharagpur, India   Dilip Kumar Pratihar

Contents

About the Authors

Dr. Abhijit Mahapatra received B.E. and M.Tech. degrees in Mechanical Engineering from B.E. College (now, BESU), Shibpur, India, and NIT Durgapur, India, in 2002 and 2008, respectively. He received his Ph.D. from NIT Durgapur, India, in 2018. He is currently working as a Principal Scientist in the Advanced Design and Analysis Group at CSIR- Central Mechanical Engineering Research Institute, Durgapur, India.

Dr. Mahapatra has published a number of research papers in national and international journals and conference proceedings, and has filed a number of patents in the area of product development. His current research interests include design & analysis, multi-body dynamics, and modelling and simulation of legged robots.

Dr. Shibendu Shekhar Roy received B.E. and M.Tech. degrees in Mechanical Engineering from NIT, Durgapur. He also holds a Ph.D. from IIT, Kharagpur, India. He is currently working as a Professor at the Department of Mechanical Engineering and Associate Dean (Alumni Affairs & Outreach) at the National Institute of Technology, Durgapur.

Dr. Roy has published more than 68 papers in national and international journals and conference proceedings, as well as 4 book chapters, and has filed a number of patents in the area of product development. His current research interests include modelling and simulating legged robots, soft robotics, rehabilitation robotics, additive manufacturing and 3D printing on macro- and micro-scales.

Dr. Dilip Kumar Pratihar completed his B.E. and M. Tech. in Mechanical Engineering at NIT, Durgapur, India, in 1988 and 1995, respectively. He received his Ph.D. from IIT Kanpur in 2000. Dr. Pratihar pursued postdoctoral studies in Japan and then in Germany under the Alexander von Humboldt Fellowship Program. He is currently working as a Professor at IIT Kharagpur, India. His research areas include robotics, soft computing and manufacturing science.

He has made significant contributions in the development of intelligent autonomous systems in various fields, including robotics, and manufacturing science. He has published more than 230 papers, mostly in international journals, and is on the editorial board of 12 international journals. He is a member of the FIE, MASME and SMIEEE. He has completed a number of sponsored (funded by DST, DAE, MHRD, DBT) and consultancy projects and is a member of Expert Committee of Advanced Manufacturing Technology, DST, Government of India.

Nomenclature

Latin Symbols

$\mathbf{A}^{G_0 G}$	3×3 Transformation matrix that maps *dynamic* reference frame ($\boldsymbol{G}$) into *static* global reference frame ($\boldsymbol{G_0}$)
$\mathbf{A}^{G_0 N_{i3}}$	3×3 Transformation matrix that maps reference frame ($\boldsymbol{N_{i3}}$) into reference frame ($\boldsymbol{G_0}$)
$\mathbf{A}^{L_0 G}$	3×3 Transformation matrix that maps *dynamic* reference frame ($\boldsymbol{G}$) into TB body reference frame ($\boldsymbol{L_0}$)
$\mathbf{A}^{L_0 L'_{ij}}$	3×3 Transformation matrix that maps reference frame ($\boldsymbol{L'_{ij}}$) into reference frame ($\boldsymbol{L_0}$)
$\mathbf{A}^{L''_{i3} L_{i3}}$, $\mathbf{A}^{L'''_{i} L_{i}}$	Transformation matrices that map reference frames ($\boldsymbol{L_{i3}}$) into reference frame ($\boldsymbol{L''_{i3}}$) and ($\boldsymbol{L'''_{i3}}$) respectively
$\mathbf{A}_i$	is a square matrix, $\in \mathbb{R}^{3,3}$
$\mathbf{A}_e$, $\mathbf{B}_e$, $\mathbf{A}_u$, $\mathbf{B}_u$	are combined matrices
$\mathbf{B}_t$	is a square matrix, $\in \mathbb{R}^{3,3}$
c	An auxiliary variable, $\in \mathbb{R}^{18}$
$\mathbf{C}_B$, $\mathbf{C}_L$, $\mathbf{C}_T$	COM of the trunk body, COM of the payload, aggregate center of mass of the system respectively
$\mathbf{C}_k$	Center of mass of the rigid body 'k' of the hexapod (here, $k = 0$ designates combination of trunk body and payload; $k = ij$ designates link ij, for $i = 1$ to 6, $j = 1$ to 3)
$\mathbf{C}_i$	COM of the interactive contact volume of the foot tip with the terrain
$\mathbf{C}p_i$	COM of footpad

C_{max}	Maximum damping
$\mathbf{D}_i$	is a square matrix, $\in \mathbb{R}^{3,3}$
$\mathbf{D}(q)$	Coupled mass and inertia matrix of the system in terms of generalized coordinates
d	Diameter of footpad
d_{ij}	Joint offset distance
e	Force exponent (mostly material property and $e > 1$ for stiff spring)
E_s	Specific energy consumption
$\mathbf{f}_K, \mathbf{f}_U$	Total summation of known and unknown forces and moments, respectively
$^c\mathbf{f}$	Vector of constraint reaction forces and torques of the joints in terms of Cartesian coordinates, $\in \mathbb{R}^{114}$
$\mathbf{f}(\mathbf{p}^{G_0}, \mathbf{v}^{G_0})$	Vector of applied forces and torques
$\mathbf{F}_i^G, \mathbf{F}_i^{G_0}$	Vector of the ground reaction forces at the foot of leg i with respect to G and G_0, respectively
$\mathbf{F}_e^{G_0}$	Vectors representing the centrifugal, Coriolis, gyroscopic, and gravitational forces acting on the combined mass of the trunk body and payload with respect to frame G_0
$^m\mathbf{F}_i^G$	Vector of resultant compliant impact force with respect to frame G
$\mathbf{F}_{C_m}^{G_0}$	Vector of gravitational forces acting on the system with respect to frame G_0
G	*Dynamic* reference frame corresponding to XYZ axes system
G_o	*Static* global reference frame corresponding to XYZ axes system
g	Acceleration due to gravity
$\bar{\mathbf{H}}$	An auxiliary variable known as Hessian matrix, $\in \mathbb{R}^{18 \times 18}$
$\mathbf{H}^{G_0}$	Total angular momentum of the system expressed in G_0
h_c	Depth of the centroid of the interactive contact volume from the plane XY
h'_{in}	Height of the terrain at the point of touch by swing leg i with the ground at nth cycle
h_l	Vertical height that gives the measure of energy stability level
Hm_{in}	Maximum height of the topography on the path of the trajectory of swing leg i at nth cycle

Δh	Clearance between the maximum height of trajectory (along Z-direction) of swing leg i and Hm_{in} at nth cycle
i	Leg number (here $i = 1$ to 6)
j	Joint number (here $j = 1$ to 3)
$J_{C_k x}^L,\ J_{C_k y}^L,\ J_{C_k z}^L,\ J_{C_k xy}^L,\ J_{C_k yz}^L,\ J_{C_k zx}^L$	Mass moment of inertia of the rigid body 'k' of the system about x, y, z, xy, yz, and zx with respect to frame L_0 respectively [here, $L = L_0$ for $k = 0$; $L = L'_{ij}$ for $k = ij$]
$\mathbf{J}_{C_k}^L,\ \mathbf{J}_{P_k}^L$	Mass moment of inertia matrix of the rigid body 'k' of the system with respect to the points $\mathbf{C}_k$ and $\mathbf{P}_k$, respectively, represented in the local (body-fixed) frame L [here, $L = L_0$ for $k = 0$; $L = L'_{ij}$ for $k = ij$]
$\mathbf{J},\ \dot{\mathbf{J}}$	Jacobian matrix and its derivative, respectively, for the kinematically transformed system in terms of generalized coordinates
$\mathbf{J}_{r_{ij}},\ \mathbf{J}'_{r_{ij}}$	3×6 Jacobian matrices of the *inverse kinematics*
$\dot{\mathbf{J}}_{r_{ij}},\ \dot{\mathbf{J}}'_{r_{ij}}$	3×6 Time derivative of the Jacobian matrices
K	Characteristic stiffness of the boundary surface considered in MSC.ADAMS®
k	Body part number (here, $k = 0$ designates combined trunk body and payload; $k = ij$ designates link ij)
l	Edge or segment of support polygon considered as the rotation axis
l_i	Leg offset distance, that is, minimum projective distance between the points at $\mathbf{S}_i$ and $\mathbf{P}_{i3}$
l_{ij}	Kinematic link length of link ij
l'_{i3}	Minimum distance between rotating axis of joint $i3$ and foot-pad center
L_0	Body-fixed reference frame located at point $\mathbf{P}_0$ on the trunk body
$\mathbf{L}^{G_0}$	Total linear momentum of the system expressed in G_0
$L'_i,\ L''_i$	Reference frames on joint $i1$ indicating successive joint states respectively
$L'_{i1},\ L''_{i1}$	Reference frames on joint $i2$ indicating successive joint states respectively
$L'_{i2},\ L''_{i2}$	Reference frames on joint $i3$ indicating successive joint states respectively

L_{i3}, L'_{i3}, L''_{i3}, L'''_{i3}	Reference frames located at tip point $\mathbf{P}_{i3}$ and parallel to reference frame G; parallel to reference frame L''_{i2}; signify reference frame for crab angle during crab motion; signify reference frame for turning angle during turning motion respectively
m_B, m_L, m_{ij}	Masses of trunk body, payload, and links ij, respectively
m	Number of divisions of a gait cycle
m_b	Mobility of the system
$M_{ij}^{G_0}$	Torque at the jth joint of ith leg,
$M_{ij,\min}$ and $M_{ij,\max}$	Torque limitations at the jth joint of ith leg
$\mathbf{M}_k(\mathbf{p}_k^{G_0})$	Mass matrix of the rigid body 'k' in terms of Cartesian coordinates, $\in \mathbb{R}^{6,6}$
$\mathbf{M}(\mathbf{p}^{G_0})$	Combined mass matrix of the system in terms of Cartesian coordinates, $\in \mathbb{R}^{114,114}$
$\mathbf{M}_0^{G_0}$	Vector of joint torques acting on the trunk body and payload (combined)
$\mathbf{M}_e^{G_0}$, $\mathbf{M}_{ei}^{G_0}$	Vectors representing the centrifugal, Coriolis, gyroscopic, and gravitational moments acting on the trunk body, payload, and leg i with respect to frame G_0 respectively
$\mathbf{M}_i^{G_0}$	Vector of joint torques of leg i, $\in \mathbb{R}^{18}$
$\mathbf{M}^{G_0}$	Overall joint torque vector, $\in \mathbb{R}^{18}$
n	Number of gait cycles
n_c	Number of kinematic joint constraints
n_g	Number of grounded legs
$\mathbf{O}$	Global origin or point of coincident of the reference frames G_0 and G
p	Boundary penetration ($\ll z_1$) at which MSC.ADAMS® solver applies full damping
P_{av}	Average power consumption
$\mathbf{p}_k^{G}$	Vectors of Cartesian coordinates of point $\mathbf{P}_k$ with respect to G_0
$\mathbf{p}^{G}$	Overall vectors of Cartesian coordinates of the system with respect to G
$\mathbf{p}^{G_0}$	Overall vectors of Cartesian coordinates of the system with respect to G_0
$\mathbf{P}_k$	Point representing the origin of the local reference frame located on the rigid body 'k' of the hexapod
$\mathbf{P}_r^T(x,y)$, $\mathbf{P}_r^T(x,z)$	2×3 matrix projectors

$\mathbf{P}_r(y)$, $\mathbf{P}_r(z)$	1×3 matrix projectors
$\mathbf{q}_0$, $\mathbf{q}_i$, $\mathbf{q}$	Generalized coordinates of the trunk body, leg i and overall system respectively
$\mathbf{q}_{GC}$ ($\mathbf{p}^{G_0}$, $\mathbf{v}^{G_0}$)	Vector of centrifugal forces and gyroscopic terms
$\mathbf{Q}_i$	Friction coefficient matrix, $\in \mathbb{R}^{4,3}$
$\mathbf{r}^{L_0}_{C_L P_0}$, $\mathbf{r}^{L_0}_{C_T P_0}$, $\mathbf{r}^{L_0}_{S_i P_0}$	Displacement vector from point $\mathbf{P}_0$ to $\mathbf{C}_L$, $\mathbf{C}_T$, and $\mathbf{S}_i$, respectively represented in reference frame $\mathbf{L}_0$
$\mathbf{r}^{L'_{i1}}_{P_{i1} S_i}$, $\mathbf{r}^{L'_{i2}}_{P_{i2} P_{i1}}$, $\mathbf{r}^{L'_{i3}}_{P_{i3} P_{i2}}$	Displacement vectors from point $\mathbf{S}_i$ to $\mathbf{P}_{i1}$, $\mathbf{P}_{i1}$ to $\mathbf{P}_{i2}$, $\mathbf{P}_{i2}$ to $\mathbf{P}_{i3}$, respectively, in their local reference frames $\mathbf{L}'_{i1}$, $\mathbf{L}'_{i2}$, and $\mathbf{L}'_{i3}$
$\mathbf{r}^{L'_{ij}}_{C_{ij} P_{ij}}$	Displacement vector from point $\mathbf{P}_{ij}$ to $\mathbf{C}_{ij}$ represented in reference frame $\mathbf{L}'_{ij}$
$\mathbf{r}^G_{C_0 O}$, $\mathbf{r}^G_{C_{ij} O}$, $\mathbf{r}^G_{C_m O}$	Displacement vector from point $\mathbf{O}$ to $\mathbf{C}_0$, $\mathbf{C}_{ij}$, and $\mathbf{C}_m$ represented in reference frame $\mathbf{G}$ respectively
$\mathbf{r}^G_{P^s_{i3} O}$	Displacement vector of the tip point $\mathbf{P}_{i3}$ of leg i at any instant of time
$\mathbf{r}^G_{P_0 O}$	Displacement vectors represented in reference frame $\mathbf{G}$ from point $\mathbf{O}$ to $\mathbf{P}_0$
$\mathbf{r}^G_{P_{ij} O}$	Displacement vectors represented in dynamic reference frame $\mathbf{G}$ from point $\mathbf{O}$ to $\mathbf{P}_{ij}$ (for $j = 1$ to 3)
$\mathbf{r}^G_{S_i O}$	Displacement vectors represented in dynamic reference frame $\mathbf{G}$ from point $\mathbf{O}$ to $\mathbf{S}_i$
$\dot{\mathbf{r}}^G_{P_0 O}$, $\ddot{\mathbf{r}}^G_{P_0 O}$	Translational velocity and acceleration vectors of point $\mathbf{P}_0$ respectively, represented in reference frame $\mathbf{G}$
$\dot{\mathbf{r}}^G_{P_{ij} O}$, $\ddot{\mathbf{r}}^G_{P_{ij} O}$	Translational velocity and acceleration vectors of point $\mathbf{P}_{ij}$ respectively, represented in reference frame $\mathbf{G}$
$\mathbf{r}^{G_0}_{C_m O}$	Displacement vector from point $\mathbf{O}$ to $\mathbf{C}_m$ represented in reference frame $\mathbf{G}_0$
$\mathbf{r}^{G_0}_{C_m/l}$	Vector from line $\mathbf{P}_{13} \mathbf{P}_{53}$ to COG with respect to frame $\mathbf{G}_0$ and is orthogonal to line $\mathbf{P}_{13} \mathbf{P}_{53}$
$\mathbf{r}'^{G_0}_{C_m/l}$	Vector obtained by rotating vector $\mathbf{r}^{G_0}_{C_m/l}$ about line $\mathbf{P}_{13} \mathbf{P}_{53}$, until it lies in plane I (vertical plane which includes the line $\mathbf{P}_{13} \mathbf{P}_{53}$, that is edge l)
$\dot{\mathbf{r}}^{G_0}_{P_{ij} O}$, $\ddot{\mathbf{r}}^{G_0}_{P_{ij} O}$	Translational velocity and acceleration vectors of point $\mathbf{P}_{ij}$ respectively, represented in reference frame $\mathbf{G}_0$

$\mathbf{S}_i$	Origin or point of coincident of the reference frames $\mathbf{L'}_i$ and $\mathbf{L''}_i$ associated with joint $i1$
$s_0,\ s_0^c$	Linear and angular stroke of the trunk body respectively
s_T	Total distance travelled during locomotion
s_w^f	Full swing stroke of Leg i
s_0''	Distance covered by the trunk body per division per duty cycle
t_0	Start time of the trunk body motion
t_1	Time at which velocity of trunk body reaches maximum
t_2	Time at which the trunk body starts decelerating
t_3	End time of the trunk body motion
$t_3^s\big|_n$	End time of the swing leg motion for the nth cycle
t_{start}^s	Start time of the swing leg motion
t_{end}^s	End time of the swing leg motion
$\mathbf{T}_i^{G_0}$	Vectors of moments at the foot of leg i with respect to G_0
$\mathbf{T}_{C_m}^{G_0}$	Resultant moment acting at COG
$\mathbf{u}$	Velocity vector of the overall system in terms of generalized coordinates
$\dot{\mathbf{u}}$	Acceleration vector of the overall system in terms of generalized coordinates
$v_{xy}^{N_{i3}}$	Slip velocity
v_s	Stiction transition velocity
v_d	Friction transition velocity
$\mathbf{v}_k^{G_0}$	Velocity vector of the rigid body 'k' of the system measured with respect to G_0 in the Cartesian system
v_{ref}	Reference velocity
v_{t_0}	Initial rate of change of displacement (linear or angular) of the trunk body with respect to point $\mathbf{P}_0$ is considered to be,
v_{t_1}	Maximum rate of change of displacement (linear or angular)
$\mathbf{W}$	Symmetric positive definite matrix, $\in \mathbb{R}^{18,18}$
z	Distance function of the impact
z_1	Trigger distance (vertical) between COM of foot-pad ($\mathbf{C}p_i$) and edge of foot-pad ($\mathbf{H}$)
$\dot{z}$	Derivative of z to impact

Greek Symbols

α_G, β_G, θ_G	Angles representing slope, banking, initial position of the robot on the terrain, respectively
α_0, β_0, θ_0	Angular displacement of trunk body along X-, Y-, and Z- directions with respect to frame L_0, respectively
θ_{i1}, β_{i2}, β_{i3}	Angular displacement of joint $i1$ located at S_i, joint $i2$ located at P_{i1}, joint $i3$ located at P_{i2}, respectively
γ	Fixed angle between frames L_0 and L'_i
ϕ	Twisted angle of the coxa
ψ	Bounding angle between l_{i3} and l'_{i3}
Ψ_l	Angle between $\mathbf{r}'^{G_0}_{C_m/l}$ and unit vector $\hat{\mathbf{z}}$
Ω_l	Angle between $\mathbf{r}^{G_0}_{C_m/l}$ and $\mathbf{r}'^{G_0}_{C_m/l}$
θ_c	Crab angle
$\theta^{L_{i3}}_{i3}$	Turning angle
ε_s	Slip angle
μ_s	Coefficient of static friction
μ_d	Coefficient of dynamic friction
ζ	Friction angle
δ_i	Angle of inclination of the flat surface of the footpad with respect to reference frame G
χ	Angle subtended by the impact force $^m\mathbf{F}^G_i$ with the point C_i
τ	Vector of ground reaction forces/ moments and coupled joint torques
λ	Vector of the Lagrange multipliers, $\in \mathbb{R}^{n_c}$
$^{G_0}\boldsymbol{\omega}_{ij}$, $^{G_0}\dot{\boldsymbol{\omega}}_{ij}$	Vector of angular velocity and angular acceleration of the link ij represented in frame G_0, respectively
Δ'_a and Δ'_d	First derivatives of Δ_a and Δ_d, respectively, with respect to time
Δt	Duration of acceleration and deceleration of the trunk body

Abbreviations

3D	Three-dimensional
ADAMS	Automated dynamic analysis of mechanical systems
CAD	Computer-aided design
CAE	Computer-aided engineering
CATIA	Computer-aided three-dimensional interactive application
COG	Center of gravity
COM	Center of mass
DF	Duty factor
DGSM	Dynamic gait stability measure
DH	Denavit–Hartenberg

DOF	Degrees of freedom
FBD	Free body diagram
NE	Newton–Euler
NESM	Normalized energy stability margin
QP	Quadratic programming
VP	Virtual prototyping
ZMP	Zero moment point

List of Figures

List of Tables

Chapter 1
Introduction

In this chapter, a brief introduction is given to the multi-legged robots walking in various terrains. Moreover, it covers the advantages and disadvantages of legged robots followed by its practical applications and methodologies involved in the modeling of multi-legged robots.

1.1 Introduction to Multi-legged Robots

Over the past two decades, the field of robotics has emerged as one of the major areas of challenge and intense research, whether it is either all-terrain robot (ATR) or autonomous underwater vehicle (AUV) or articulated robotic hand or exoskeleton robotics suit. Again, each of these sub-areas has to tackle some of the major technological challenging areas like (i) sensors, (ii) mobility and power, (iii) navigation, (iv) communication and control, and (v) autonomy. Keeping pace with time, the field of ATR has emerged as one of the interesting and provoking grounds to researchers worldwide due to its wide choice of consideration, whether it is either legged or wheeled or tracked vehicles or hybrid vehicles for locomotion. So far as high degree of terrain adaptability and maneuverability, that is, capabilities of stepping over obstacles or ditches, climbing over obstacles or maneuvering within a confined area of space is concerned, legged robots are preferred to others, since they can negotiate any kind of hostile environment. It can adjust its leg lengths to maintain its body level and leg extensions to position the body's center of gravity during travel. The traveled path consists of a series of discrete points only in contact with the terrains, not like a continuous path followed by a wheeled or tracked robot.

Today, research in robotics aims to build intelligent and adaptive walking robots to meet increasing demand for performing tasks, which are dangerous and difficult for human beings, e.g., anti-terrorist action, detection of bomb exploration, disaster

© Springer Nature Singapore Pte Ltd. 2020
A. Mahapatra et al., *Multi-body Dynamic Modeling of Multi-legged Robots*,
Cognitive Intelligence and Robotics, https://doi.org/10.1007/978-981-15-2953-5_1

salvation, planetary exploration, etc. An adaptive multi-legged robot will have to plan its collision-free path along with the gait in such situations during its locomotion to accomplice the given task. However, it is important to note that an intelligent and adaptive walking robot or autonomous walking robot cannot perform satisfactorily with a low energy efficiency due to the fact that it has to carry all driving and control units in addition to its trunk body and payload. Therefore, long duration mission is subject to power supply constraints. Hence, minimizing the energy consumption during locomotion is important for an autonomous multi-legged robot system used for service application. Reduction in energy consumption for any service application robot not only helps it to travel long but also miniaturizes the actuators leading to reduced body weight and cost.

1.2 Legged Robot's Locomotion

A gait of a multi-legged robot is defined as a sequence of legs followed during locomotion in coordination with the body's movement. Gaits are basically of two types: periodic gaits (wave gaits, equal-phase gaits, etc.) and aperiodic gaits (free gaits, adaptive gaits, etc.), as shown in Fig. 1.1. A periodic gait is suitable for a smooth and flat terrain, while an aperiodic gait is applicable to rough terrains (e.g., various obstacles including rocks, soils, sands, slanted surface, and ditches, etc.). Also, gaits can be classified as level-walking gaits and obstacle-crossing gaits, etc. [10]. Level-walking gaits can be categorized into two types: gaits according to the terrain condition (i.e., gaits on perfectly smooth terrain) and gaits on terrain that contains forbidden areas. It is to be mentioned that level-walking gaits are generally suitable for smooth terrain and do not require the body to be raised, lowered, or tilted during walking. However, if the terrain irregularities exceed a certain, limit, the body must be raised, lowered, or tilted in accordance with the legs movement.

During locomotion on various terrains, a multi-legged robot might have to move along straight paths, ascend and descend sloping terrains, take turns or move in transverse direction at certain crab angle to avoid some obstacles, and others, as the situation demands. Such movements are possible only with some of the important level-walking gaits on various terrains. A brief description of each gait is given below. Straight-forward walking is defined here as the straight walking, the direction of motion of which remains parallel to the position of the longitudinal axis of the robot's body at the start of the gait, by taking also into consideration the changes in body orientation (roll, pitch, and yaw angles) and variation in body height of the robot in various terrains. In other words, here, the angle between the direction of motion and the longitudinal axis of the robot's body is zero. Likewise, crab walking is defined as the walking in a direction other than the longitudinal axis of the robot body in a various terrains with the changes in body orientation (roll, pitch, and yaw angles) and height. The angle which governs the crab walking is called the crab angle and is defined as that between the direction of motion and the longitudinal axis of the robot body. Similarly, for turning gaits, the walking motion of the robot is considered

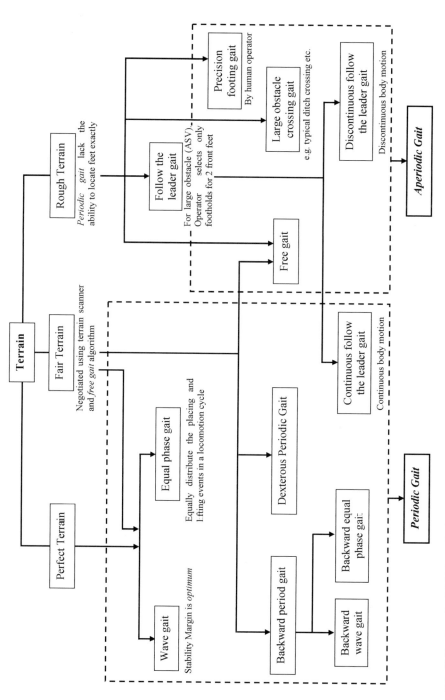

Fig. 1.1 Gait pattern for Multi-legged robots

around a turning center with the head changing direction instantaneously, as it turns. The other conditions like changes in body height, body orientation (roll, pitch, and yaw angles) are kept similar to that of other gaits mentioned above.

Gaits can be further classified on the basis of walking, namely static and dynamic gaits. Static gait is characterized by slow motion with negligible dynamic effects (neglecting the inertial forces) during locomotion, while the dynamic gaits are fast and their dynamic effects are significant taking into account the inertial forces. Moreover, the dynamic gait generation problem of a multi-legged robot working in the environment containing some sloping surface, ditch, and staircases is a challenging task. However, coordination of various leg joints so as to produce the desired gait pattern and maintain stability during locomotion is extremely complex [6]. To achieve it, development of a robust controller, which controls velocity and posture of the robot while maintaining a cyclic gait, is required. Therefore, a good mathematical model of the complex multi-legged robotic mechanism describing the kinematics and dynamic behavior is necessary, so that it will be able to address the abovementioned issues and analyze energy consumption of legged robots.

1.2.1 Leg Mechanisms and Comparisons: Multi-legged Robots

The most important feature of any multi-legged robot is the number of legs which actively take part in the locomotion. The number of legs depends on the application for which the robot is built and the variety of different features that it includes. A balance between the two needs to be studied in terms of stability, speed, reliability, weight, and overall cost. The design of the legs can be varied with the variation in the degrees of freedom. Although the configuration of each leg of a robot is similar to a manipulator, the design requirements are different due to high payload-to-weight ration. The leg designs can be broadly classified into three categories, namely (a) straight-line mechanism, (b) articulated mechanism, and (c) pantograph leg mechanism.

The straight-line mechanism of the legs corresponds to the propelling motion of the trunk body with single degree of freedom. This makes the control simpler. However, there is limitation in such type of design like poor adaptability, restricted workspace, and fixed gait.

Articulated leg mechanism has been the most popular design for multi-legged robots. The mechanism has three degrees of freedom (three rotary joints), two at the hip joint and one at the knee joint. In such type of mechanism, two of the joint actuators are mounted on the trunk body, thereby making the leg lighter and reducing the inertia effect of the swing leg on other parts. However, such mechanism has two kinds of variants: (a) insect type and (b) mammalian type. The key variation is in the placement of the knee joint. The knee joint in insect type legs is placed laterally at a height from the hip joint, whereas in mammalian type legs it is placed directly below the hip. Although both the variants have large workspace and simplified kinematics, insect leg type is desirable, because it is less likely to change joint torque signs compare to mammalian leg type [6].

The mechanical design of the legs adopting pantograph leg mechanism is spatially complex. However, the mechanism can provide straight-line motions in horizontal and vertical directions. It also helps in amplifying the hip motion at the foot. The linkage also allows the actuators to be kept close to the trunk body, thereby, reducing the inertia effects.

Generally, a statically stable multi-legged robot is used in the form of a four-legged, six-legged robot, or more. However, studies show that a six-legged robot is better suited than others for maneuverability over any kind of terrain. It has many advantages compared to two- or four-legged robots like,

- provides better static stability
- less susceptible to deadlock situations
- more robust, as it is able to walk with one or two failed legs (it is possible to define stable gaits by using either four or five legs).

Also, compared to eight- or more-legged robots, six-legged robot's power consumption is less, since with the increase in number of the legs a lot of actuators are to be controlled through a continuous coordination in addition to complicacies in kinematics and dynamics of legged mechanisms [1, 4, 5, 7–9]. Therefore, a six-legged robot is a kind of optimal robotic structure to be used for various terrains.

1.2.2 Advantages of Multi-legged Robots

Locomotion of a robot on any rough-uneven-loose terrain is a complex task and is best possible with legs only. This advantage of legged locomotion is due to the fact that legged systems use isolated footholds that optimize support and traction. Wheeled and tracked vehicles follow any terrain surface in a continuous manner, and hence their performance is limited by the degree of unevenness of the terrain. A legged system on the other hand can choose the best area on the terrain for foot placement. In other words, it is well adaptive to uneven terrains, namely the legs can be arranged (lengthened and shortened) according to the level changes, and they can jump over obstacles or holes. However, these footholds are isolated from the rest of the body, and hence its performance is limited by the best footholds. Other than this, legged systems provide active suspension, which does not exist in wheeled or tracked systems. This means that the system can decouple the path of the trunk body from the paths of the feet, which helps it to travel freely and smoothly despite pronounced variations in the terrain. Even the system can step over obstacles which means in principle, the performance of legged robot can, to a great extent, be independent of the detailed roughness of the ground. Therefore, it is to be noted that this decoupling is used by any legged system to increase its speed and efficiency on rough terrain. Again, a legged system can have control on the force distribution through the foothold points.

In this way, an efficient utilization of the footholds provides further improvement for the vehicle-ground interaction. Further, when a legged system locomotes in an environment, it causes less damage to the terrain surface compare to the wheeled and tracked systems do [6]. Consideration of all these advantages makes the application of a legged system in various fields far superior to wheeled and tracked systems.

1.2.3 Disadvantages of Multi-legged Robots

Even though there are many advantages, legged robots possess few drawbacks also which are required to be dealt with. For example, the locomotion kinematics and dynamics of legged mechanisms are complicated. A lot of actuators are to be controlled through a continuous coordination for achieving synchronized motion and fulfilling the required task. Therefore, its control is considerably tough in comparison with wheeled or tracked systems. Moreover, its energy consumption is high, in contrast to its payload-to-weight ratio which is low compared to wheel and tracked robots. Again, in multi-legged robots each of the joints is driven by a separate motor which is the heavier compare to the link. Also, the number of motors in each leg is equal to the number of degrees of freedom (DOF) which leads to a cumulative many motors in the system of multiple legs. In general scheme, the motors are placed on the joints though there are few exceptions in the design of multi-legged robots. These motor weights in the legs are quite comparable to that of the trunk body. Normally, the actuators on a leg are serially connected. Therefore, the weight a leg can carry is limited by the power of the weakest actuator on the leg. During locomotion with a specified gait (e.g., tripod gait with three legs in swing and three in support phase), the legs in swing phase do not contribute to carrying of the payload; however, there exists an extra weight which the system has to bare at the cost of energy. These factors due to the 'legs' also contribute to the low ratio of payload-to-weight of legged systems. Further, Hardarson et al. [2] also mentioned a drawback of multi-legged robot, that is, the vehicle suffers a force impact in a pulse form in every step which may result in rocking on the body. However, there are studies to overcome such impulses with proper force distribution [3]. In comparison with wheel robot, the speed of legged robot is very less in a plain, regular surface. However, the legged robot can achieve higher speeds than wheeled or tracked system on highly rough terrain. Since the development of multi-legged robot is a newly developing field, compared to wheeled and tracked systems, there are no well-established technologies for legged systems.

1.2.4 Applications of Multi-legged Robots

Multi-legged robots have an edge over tracked or wheeled robots so far locomotion over various terrains are concerned. Some of the application areas of legged locomotion are like terrestrial exploration or particle gathering from surfaces of planets including earth or moon, recoveries of objects or human beings from the regions of fire or earthquake disasters, locomotion in forests and agricultural lands, inspection of nuclear power plants, security or surveillance, safe transportation in crowded places, etc. However, in the recent past there is a lot of demand for autonomous multi-legged robots that can perform some of the hazardous tasks and substitute human interventions, e.g., anti-terrorist action, detection and defusing of bombs, hostile territory exploration, etc.

1.3 VP Tools for Modeling and Analysis of Multi-legged Robots

Over the last three decades, a lot of study has been carried out by different researchers on multi-legged robots. The four main important issues which are vividly studied include gait planning, dynamics, control, and stability (both static and dynamic) analysis of multi-legged robots. Most of the studies involved the design and development of the robot system for real experiment and subsequently validation. However, in the recent past, use of virtual prototyping (VP) tools in integrated product development like robotic system has been a current trend in research. The later approach has advantages over the first, like lower development cost, reduction in the number of real prototypes for testing, less time for design modification and prototype development, etc. Based on such VP tools, several simulation models can be developed and used for study, design optimization, stability (dynamic and static), gait analysis, development of control algorithms, etc. The steps involved in the later approach are summarized in Fig. 1.2.

The book consists of five chapters. The contents of each chapter are described below in brief.

Chapter 2 reviews the variety of aspects of multi-legged locomotion which are basically the kinematics, dynamics, foot-terrain interaction, energy efficiency and stability analysis of multi-legged robots.

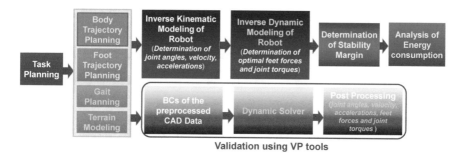

Fig. 1.2 Methodology in modeling and analysis of multi-legged robots using VP tools

Chapter 3 essentially covers the generalized kinematic formulations of the problems related to straight-forward, crab, and turning gaits of the six-legged robot on various terrains. Section 3.1 gives a description of the present problem along with gait terminologies. Section 3.2 with sub-sections describes a generalized analytical framework that involves methodology to solve kinematics of the system in various terrains with proposed path planning and gait planning. Section 3.3 describes some case studies to test the performance of the model through computer simulations.

Chapter 4 formulates the problem for coupled non-linear dynamic systems and analyzed the optimal feet forces, energy consumption, and stability of the robot in various terrains. Section 3.1 is related to the analytical formulation of the coupled dynamical model with foot-ground interaction mechanics, optimal feet force distribution, and stability analysis of the robotic system in various terrains. Thereafter, Sect. 3.2 illustrates some case studies to test the effectiveness of the developed model through computer simulations.

Chapter 5 validates the developed analytical model with that of VP tools and experiments. The aim of this chapter is to prove the robustness of the developed analytical formulation and provide theoretical basis for developing algorithms for robot motion control.

1.4   Summary

This chapter provides a brief introduction to multi-legged robots, its advantages and disadvantages, practical applications, and methodology to be followed in modeling and analysis of multi-legged robots. Finally, it provides the organization of this book.

References

1. M. Akdag, H. Karagülle, L. Malgaca, An integrated approach for simulation of mechatronic systems applied to a hexapod robot. Math. Comput. Simul. **82**, 818–835 (2012)
2. F. Hardarson, T. Benjelloun, T. Wadden, J. Wikander, *Biologically inspired design and construction of a compliant robot leg for dynamic walking*. in Proceedings of the 6th UK Mechatronics Forum International Conference (Skovde, 1998) pp. 367–372
3. L.S. Martins-Filho, R. Prajoux, Locomotion control of a four-legged robot embedding real-time reasoning in the force distribution. Robot. Auton. Syst. **32**, 219–235 (2000)
4. X.Y. Sandoval-Castro, M. Garcia-Murillo, L.A. Perez-Resendiz, E. Castillo-Castañeda, Kinematics of Hex-Piderix—a six-legged robot—using screw theory. Int. J. Adv. Rob. Syst. **10**(19), 1–8 (2013)
5. P.G. De Santos, E. Garcia, J. Estremera, *Quadrupedal Locomotion: An Introduction to The Control of Four-Legged Robots* (Springer-Verlag Limited, London, 2006)
6. S.M. Song, K.J. Waldron, *Machines That Walk: The Adaptive Suspension Vehicle* (The MIT Press, Cambridge, 1989)
7. S.S. Roy, D.K. Pratihar, Effects of turning gait parameters on energy consumption and stability of a six-legged walking robot. Robot. Auton. Syst. **60**(1), 72–82 (2012)

8. S.S. Roy, D.K. Pratihar, *Modeling and Analysis of Six-legged Robots* (Lap Lambert Academic Publishing, 2012b)
9. S.S. Roy, D.K. Pratihar, Kinematics, dynamics and power consumption analyses for turning motion of a six-legged robot. J. Intell. Rob. Syst. **74**(3–4), 663–668 (2014)
10. C.D. Zhang, S.M. Song, Stability analysis of wave-crab gaits of a quadruped. J. Robo. Syst. **7**(2), 243–276 (1990)

Chapter 2
Multi-Legged Robots—A Review

In this chapter a comprehensive literature survey has been carried out for multi-legged robots. Different aspects of multi-legged robot locomotion pertaining to kinematic, dynamics, foot-terrain interaction, energy efficiency and stability are discussed. The survey has resulted into the identification of gaps in the literature. This is followed by aims and objectives of the present work and notable contributions made in this study.

2.1 Gait Planning of Multi-Legged Robots

As mentioned in previous chapter, there are two primary reasons to employ legs (rather than wheels) for locomotion, such as (a) discrete foothold and (b) rapid traversal over a variety of mild obstacles. But still, the full fledged application of legged robots for service purposes is yet to come in the limelight. The reason for such poor state of the art of multi-legged robot technology is mainly due to its poor energy efficiency. Moreover, walking robots are indeed much more complex than expected, not only due to mechanism complexities of the formed closed kinematic loops but also adding to it are electronic systems and control algorithms for this kinematic chains. The situation becomes more complex with the increase in number of legs and degrees of freedom (DOF) of each leg as discussed in previous chapter. All these issues address to one basic problem, that is, close coordination among constituent elements of the robotic system. Such coordination occurs in two levels: (a) local coordination, which involves control of individual joints of a leg to achieve desired tip control and (b) global coordination, which deals with several chains of multiple legs. Both these types of coordination involve the development of gait control algorithms, thereby, leading to the development of good mathematical models that could describe the kinematics and dynamics of the complex multi-legged robotic mechanisms. A considerable amount of work had been carried out by various investigators on these areas for multi-legged robot systems [11, 37, 180, 181].

© Springer Nature Singapore Pte Ltd. 2020
A. Mahapatra et al., *Multi-body Dynamic Modeling of Multi-legged Robots*,
Cognitive Intelligence and Robotics, https://doi.org/10.1007/978-981-15-2953-5_2

2.1.1   Kinematics of Multi-Legged Robots

Most of the previous studies on kinematics were based on the consideration that legged robots are parallel mechanism systems [177, 196]. However, the kinematics of legged robots are far more complicated compared to the parallel mechanism robots like Gough-Stewart platform mechanism etc. [4, 30, 75, 76, 87, 133, 137, 193–195] due to added number of DOF in the system [101]. Moreover, the foot placement and lifting of a leg in a walking robot change the total topology of the mechanism. Also, due to more number of driven joints, the control of a legged robot is much more complex than that of parallel robots. The developed kinematics models for the multi-legged robots by some other researchers were simplified with the assumptions like the steady-state condition of the trunk body for different speeds, zero pitching and rolling angles of the trunk body, no change in the leg's position (i.e., no slippage) in support phase [43, 62, 188]. A few researchers tried to investigate the kinematics of legged robots using screw theory and vector algebra [69, 181], while others modeled multi-legged walking robot using Denavit-Hartenberg parameters [33] to derive kinematics equations. Further, most of the studies tried to develop the kinematic models of legged robot by treating trunk body, swing legs and support legs as separate entities; [5, 11, 13, 115, 141, 155, 158, 159, 161, 162]. However, that could not depict the actual motion behavior of the robotic system. If the motion of the legged robots is closely studied, it reveals that the movement of a swing leg depends not only on the motion of its own joints, but also on the movement of support legs and angular motions (roll, pitch, yaw motions) of the trunk body in 3D Cartesian space. Therefore, dependency of all the legs with respect to each other cannot be avoided while developing the kinematic model of the system. In addition, synchronization of the legs movement in the system has to be dealt with proper gait sequencing as discussed in Sect. 1.2 [121]. If the kinematics of the system is not correctly analyzed, it will largely affect both the stability as well as dynamics of the system. More recently, researches have showed the study of radially symmetric six-legged robots [19, 107, 181], a move from the approach to consider rectangular six-legged robots. Some more studies on kinematics of multi-legged robots have also been reported recently [49, 166]. However, in all such studies, the kinematic models were simplified by neglecting one or the other issues. Those studies carried out by various investigators were focused mainly on the design and gait generation of a six-legged robot on flat terrain. In addition to that, the developed control algorithms were mainly suitable for locomotion over flat terrain, although some efforts had been made for uneven terrain locomotion [5, 111, 171]. Also, none of those models was suitable enough to be used in various terrains. Moreover, the kinematic models were developed for some specific kinds of motion like straight-forward or crab [40, 191] or turning motion [65, 130, 140] etc. on flat terrain. However, with the advance of technology, there is an ample scope to expand the robot's capabilities and achieve the kinds of real tasks for which they are suitable. Many of these tasks require trajectory tracking [43, 62, 158, 182, 185] in 3D Cartesian space, but the research conducted in this direction is not very comprehensive. In other words, the kinematic model should be versatile enough to mimic all issues related to trunk body movement, motion planning, gait and foot trajectory planning of a real robot in any kind of spatial environment. Subsequently, the kinematic model should also provide aid to the constrained dynamic model of a realistic six-legged robot to study the realistic foot-ground interaction, power consumption, dynamic gait stability of the system etc. Most of the previous models were unable to emphasize upon these issues due to considerable simplifications assumed in the kinematic models, as discussed above.

2.1.2 Dynamics of Multi-Legged Robots

Legged robots as parallel leg mechanism with various types of configurations like two- [169], four- [179], six- [143, 145], eight- [174] and multi-legged were studied in the past depending on the application, environment, speed and stability requirements. Most of those models were simplified to avoid the complexities involved in the robot dynamics. The model capabilities were limited to the study of the dynamics of the system locomoting on flat terrain [80, 143], though a very few models in the recent past were developed for locomotion on uneven terrain [145, 156]. Further, the studies were based on single leg dynamics [135] or trunk body and legs as separate entities or in some cases the effects of mass and inertia of legs were neglected [140, 143, 149, 158]. Different methods were used for formulation of the dynamics of legged robot systems like Newton–Euler method [37, 74, 178], Lagrange's method [74, 143], Kane's method [84, 138], Featherstone algorithm [44], Variational methods [10] and others. In this regard, some notable contributions by researchers worldwide are worth mentioning.

Pfeiffer et al. [129] investigated the dynamics of a stick insect walking on flat terrain. Freeman and Orin [46] studied an efficient dynamic simulation of a four-legged robot using a decoupled tree-structure approach. Lin and Song [109] developed a dynamic model of walking robots to study the dynamic stability and energy efficiency during walking along a straight path. Further, a free-body diagram approach to derive the dynamic equation based on Newton-Euler method for a six-legged walking robot was developed by Lin and Song [109]; Barreto et al. [11]. Bennani and Giri [12] modeled multi-legged robotic system using a unified dynamic modeling approach, which is based on the use of Newton-Euler formulation and explicit formulation of kinematic holonomic constraints for the closed-loop mechanisms. Based on work of Huang and Zhao [71] and a suitable dynamic manipulability ellipsoid, Zhao et al. [198] proposed a systematic method to perform dynamic analysis in the task space for six-legged robots. Silva et al. [158, 160, 161, 163] developed dynamic model of a six-legged robot with segmented trunk body and joints at the legs. Moreover the studies of some other researchers [15, 70, 127, 176] using various other methods of dynamic formulation for legged robots are worth mentioning.

The dynamic models of realistic legged robots are highly non-linear. In any realistic legged robots, the coupling effects and non-linearity in the dynamics of the swing legs on the support legs and main body are essential and should be taken into account. In addition to this, non-linearity in the interaction dynamics of the robot with the ground should also be considered in the model, though mimicking a real situation is difficult. The preceding section briefly discusses on this issue.

2.1.3 Foot-Ground Contact Modeling

Foot-ground interaction mechanics is of considerable importance in legged robots maneuvering in various terrains and has been in focus since the last decade. The magnitude of contact force has significant effect on its dynamic behavior and subsequently, upon the controller design. Therefore, modeling of contact between the foot and ground, that is, an appropriate contact force model is essential to properly investigate the dynamic behavior of a realistic robot. However, for all types of available contact force models, Hertzian contact theory [153] remains the basic model to begin with for mechanical systems.

The study of contact dynamics in mechanical systems revolves, commonly around three fields, which are as follows: (i) consideration of friction at the contact surface, that is, tangential component of contact force varies, while its normal component is assumed to be constant. It is limited to contacts with negligible normal motion and simplified geometry [24, 32, 61, 98, 152]; (ii) consideration of a frictionless contact surface i.e. tangential component of contact force is assumed to be zero, while its normal component varies. It is limited to collision of multi-body systems with perfectly frictionless contact surfaces [73, 90] (iii) general contact model by considering both tangential and normal components of the forces at the contact and taking into account impact and friction in contact interface like that found in the studies of Han and Gilmore [60], Yang et al. [183], Gonthier et al. [55], Gaul and Mayer [52], Rajaei and Ahmadian [136]. Han and Gilmore [60] developed an impact model by considering normal and tangential components of the contact force, and static and dynamic coefficient of friction at the contact interface. Yang et al. [183] investigated the effect of normal force on hysteresis loop that could describe the full-slip or full-stick condition. Gonthier et al. [55] generalized contact model included normal compliance, energy dissipation and friction force using seven parameters, namely, μ_S, μ_C, σ_o, σ_1, σ_2, v_S and τ_{dw}, where μ_S is the stiction coefficient, μ_C is the Coulomb friction coefficient (sliding), σ_o is the stiffness, σ_1 is a damping coefficient, σ_2 is the viscous damping coefficient, v_S is the velocity at which the Stribeck effect occurs, τ_{dw} is the dwell-time dynamics time constant. Gaul and Meyer [52] modeled impact force between the contact surfaces adopting a non-linear stiffness in the model. Rajaei and Ahmadian [136] generalized the Iwan model to take into account the effect of normal load variation in the contact along with friction in the tangential direction.

Legged robots are also multi-body dynamical systems, where the contact models must take into account the events like impact [167, 168, 186] and frictional contact mechanics [60] of leg-tip with the ground for successful design and optimization of the forces, torques etc. [21, 144, 164]. Both the events couple the normal and tangential (a discontinuous function of sliding velocity and independent of tangential displacement) contact forces, thereby, playing a fundamental role in the design, performance analysis, simulation and control of legged robots. Every time a leg-tip touches the ground with a non-zero velocity, impact phenomenon occurs. This phenomenon defines the transition between the swing motion of the legs and contact state characteristics, thereby imposing kinematic constraints on the foot [14]. Gradually the impact transforms into a frictional contact force problem [81, 151, 199] with slippage of the leg-tip on the terrain. When the contact points are in sliding phase, it gives rise to tangential velocities on the impact plane. However, in real life, situation may arise when any contact point can undergo sticking phenomenon. When the contact point sticks, it restricts the motion of the point and a stick-slip transition phase occurs, leading to two possibilities: stick or reverse slip [35, 167]. Such contact models can be classified into two types: rigid and compliant.

Most of the previous models dealing with contact force distribution in multi-legged robots considered the phenomenon as a rigid-body point contact (only force components) without impact and any slippage, thereby did not analyze the actual contact under each foot [26, 37, 57, 80, 91, 139, 142]. However, it is to be noted that rigid-body point contact fails to capture the full range of contact phenomena like the deformation of the leg-tip during the impact landing. Hence, a few researchers considered the leg as a compliant multi-articular structure [16, 56, 59] as discussed above. A few other researchers [23, 53] considered compliant at leg-tip and studied explicitly the normal contact force models, which are expressed as the functions of local position along normal and its rate. This compliance in the model at the interaction zone between foot and ground is necessary to absorb the unexpected impacts. Marhefka and Orin [113]; Goldsmith [54] and Ahmadi and Buehler [1] modeled the compliance as linear spring-damper with one spring. The contact model was simple with discontinuity at the initial time of impact due to the form of the damping term and sticking effect at the time of separation, holding the objects together. Featherstone [41] dealt with such kind of sticking effect by adding a state variable. Bibalan and Featherstone [17], studied the contact phenomenon with linear spring-damper considering two springs, which is a bit improved model since the contact force is continuous. Such compliant contact phenomenon was first successfully described using a non-linear spring damper visco-elastic model at the contact point of a sphere and the ground by Hunt and Crossley [73]. The model is well-known and is generally called the compliant normal contact force model which had also been described in the studies of Marhefka and Orin [113], Zhang and Sharf [197], Lankarani and Nikravesh [99], Ding et al. [34], Song et al. [164]; Vasilopoulos et al. [173]. The model was further modified by Lee and Wang [103], Gonthier et al. [55], Azad and Featherstone [9] through modifications of a single exponent following a different line of reasoning. Silva et al. [158, 160, 161, 163] modeled the foot-ground interactions dynamics of a six-legged robot using non-linear spring damper system at the tip and subsequently, developed an algorithm for foot force feedback control systems.

Besides the compliant normal contact force, there is also lateral or tangential contact force (also called friction force), which appears during the contact as mentioned above [7, 93, 123]; Gonthier et al. [57]. The most widely employed friction force model is the simple classical coulomb's (or Amonton's) model [8, 89, 172]. However, its mathematical properties complicate the dynamic simulation for both the constrained rigid body dynamics and compliant contact models. Therefore, gradually the model was further modified with due attention to presliding hysteresis and time-lag effects like the Dahl's model [32], Lugre's model [19, 24, 36], Lueven model [98, 170], GMS model [2, 18], Iwan model [78, 136]. Such models were considered with the goal of minimizing algorithm complexities and simulation time. Whatsoever, together with the compliant normal contact force model, it overcomes the discontinuity in contact force with time, handle impacts, micro-slip, hysteresis and local adhesion [7]. Further, upon the interaction of a foot with a deformable terrain, three contact forces and moments each are generated due to deformation of the contact area unlike hard point contact, where only three contact force components are generated. Though a few studies illustrated the idea of contact moments [36], its effects on the dynamics of the legged system were neglected. A better approach for tackling the real foot-ground contact situation is very essential. The present study proposes a foot-ground contact model with slip for a hexapedal robotic system, taking into consideration the visco-elastic compliant normal contact force model and velocity-based coulomb's frictional contact force model at the interaction area; which means generation of three forces and three moments on the foot. No such study on a hexapedal robotic system has been found in the literature.

2.2 Power Consumption Analysis of Multi-Legged Robots

Energy consumption is one of the major issues till date for autonomous multi-legged robots. An autonomous robot is unable to function satisfactorily with low energy efficiency, since it has to transport the payload along with the trunk body and drive control units. It is important to note that long duration missions, such as planetary exploration, under-ground mining, location and deactivation of bombs are subjected to power supply constraints. A reduction in power consumption by such robots will not only help to travel longer distances, but also reduce the size of the actuators, leading to reduction in weight and cost. Therefore, in the present state of the art, various aspect need to be improved and optimized [157]. To determine its efficiency, various researchers had adopted different approaches. Although such attempts could find the optimal values of feet forces for the multi-legged robot, they might not be suitable for real-time implementations. The reason behind is that we do not yet have approaches, which are general and flexible enough to cope with different types of irregular uneven terrains through which a robot would have to maneuver [82]. Thus, there is still a need for the development of an energy efficient and computationally tractable approach for estimation of optimized feet forces and torques.

In dynamics of legged robots with active force control, the criticality lies in the successful transmission of the body forces to the feet in contact with the ground. For static stability of multi-legged robots, minimum of three legs should be in contact with the ground at any instant. Now, if the reaction force vector on each grounded foot has three components, the feet force distribution problem becomes indeterminate. Hence, the system is redundant, that is, the solution is not unique because the number of degrees of freedom never equals the number of functional constraints. Therefore, infinite number of solutions exists for the interactive forces and moments in a constrained inverse dynamics problem subjected to additional contact constraints. In general, such redundancy is resolved using optimization methodologies. The methodologies are adopted to optimize various performance criteria and find the optimal solution corresponding to (a) foot force distribution [72, 85, 112], (b) selection of gait parameters [85, 114, 135], (c) energy efficient mechanical structure [64, 165], (d) torque distribution in joints [37, 142], (e) slippage of leg-tip or (f) deployment of energy storage devices to recover energy [3, 154] etc. The most practical performance objectives are like minimization of the sum of the squares of the ground reaction forces (naturalistic assumption according to report of Full and Tu [45]; minimization of the sum of the squares of the joint torques since heat energy loss at each joint is proportional to the square of the joint torques; minimization of power consumption for increased efficiency; minimization of tangential contact forces to avoid foot slippage etc. There are also several other alternate objective functions, which can be formulated and implemented. They are like minimization of internal force, minimization of load balance, maximization of safety margins on friction constraints, maximization of walking velocity etc.

In most of the proposed cases, the objectives are achieved using optimization methods that are mainly based on Linear Programming (LP) method [88, 92, 125], Compact-Dual LP (CDLP) Method [26, 28], pseudo inverse [50, 79, 106, 142], Quadratic Programming (QP) [37, 86, 108, 115, 120], compact QP method [27, 29], analytical methods [20, 51, 94, 110, 127], and others. These methods are evaluated to find the optimal solution for force distribution and subsequently, a real time implementation on the development of control algorithms for legged robots. A more vivid description of some of the notable work in the field of study is mentioned in the text to follow.

In the study of foot forces' distribution in support legs of walking robot, Klein and Kittivatcharapong [91] used LP method to achieve the objective of sum of the squares of the tip point force components of the robot legs and conclude that application of such methods in real time is difficult, since it leads to discontinuity in solution. Arikawa and Hirose [6] addressed the relationship between power consumption and walking gait for a quadruped robot using a simplified model (leg dynamics was not taken into account). Here, computation of the feet forces in the legs in contact with the terrain were achieved using two objectives: at first, by maximizing the walking velocity and using LP algorithm; and secondly, by minimizing power consumption and using non-linear optimization algorithm called Davidon-Frecher-Powell method. Similarly, Cheng and Orin [28] used CDLP method to optimize the foot force distribution. This method could make real time solution of the LP problem feasible, but like Klein and Kittivatcharapong [91], they were unable to overcome the discontinuity in solution. Likewise, pseudo inverse method had been utilized by many researchers (like, [109, 122, 132, 148]) to solve the feet forces' distribution problem. Roy et al. [143] developed two approaches to estimate optimal feet force distribution and joint torques of a six-legged robot for straight motion using pseudo inverse method. In that study, the foot-ground contact had been modeled as hard point contact, which is far from being realistic consideration. Moreover, the focus was on study of gaits over flat terrain. The pseudo inverse method can be used as an alternative to simplify LP method for the problem at hand, though the solution is an approximation. Moreover, the gradual distribution of forces that occurs in real robots when lifting or placing the foot was not possible with such technique. Though such method was able to provide feasible solutions to the indeterminate equilibrium equation related to forces and moments equations; equality [109, 132] and inequality constraints [46], it was not efficient for more complex problems. So, more computationally efficient methods like QP were opted by many researchers. It is suitable for large problems and produces continuous solutions.

Nahon and Angeles [120] presented the force distribution problem in mechanical fingers as a QP problem in their study of grasping. They solved the constrained optimization problem with the objective of minimizing the internal forces or norm of joint torques. It was seen to yield continuous and faster solutions compared to the LP or pseudo inverse method. Based on QP method, Chen et al. [27] developed a new approach by combining it with transformed friction constraints for optimal force distribution of quadruped robots. Further, Cheng et al. [29] and Chen et al. [26] combined the merits of compact formulation and QP to develop the compact-QP method for solving force distribution problem. The optimization algorithm could accept both linear and quadratic objective functions, such as minimum force, minimum load balance, maximum safety margin on friction constraints. The studies by Marhefka and Orin [115] showed the detailing of different approaches like internal force minimization, weighted norm of joint torque minimization, and power minimization using QP method to optimize the force distribution in the robot legs and subsequently, minimize the power supplied to each of the joint actuators. It was inferred that minimization of the sum of squares of the joint torques produces satisfactory performance considering real power dissipation in comparison with minimization of internal forces that often produces poor power performance. Similarly, Erden and Leblebicioglu [37] studied the torque distribution in six-legged robots by formulating the objective function as *minimization of the sum of the squares of the joint-torques*, and *minimization of the sum of the squares of the tip point force components* using QP method. The authors utilized modified simplex method along with Lemke's Complementary Pivoting Algorithm to compute optimum foot force and torque distributions. It was concluded from the simulation results that the former approach could give better energy efficient distribution in comparison to the later. The results of the above studies were important in the current context of the work for justification of formulating the objective function as *minimization of the sum of the squares of the joint torques* for minimum energy consumption.

Kar et al. [86] determined the feet forces' distributions in the legs of a six-legged robot with the help of sequential QP method using locomotion performance objectives like minimization of energy consumption of a walking robot. Though the solutions of feet forces are optimal, they might not be suitable for real time implementation due to the iterative nature of the method. Besides feet forces' distributions, they performed an energy efficiency analysis with respect to structural parameters, interaction forces, friction coefficient and duty factor of wave gaits, based on a simplified model of six-legged robot. It was observed that power consumption decreases with the increase in the number of supporting legs. Analysis was also carried out to study the effects of lateral offset, coefficient of friction at foot-ground interface, duty factor on energy consumption. However, such studies were carried out on simplified model and flat terrain. It is to be noted that some studies like that of Kar et al. [86] and Lin and Song [109] considered instantaneous power consumption of the model to be the product of instantaneous joint torques and joint velocities. Such models did not take into consideration the energy, which might be dissipated through each of the motors governing the motion of each joint. For that, the objective of minimizing the sum of the squares of the joint torques could be considered as a criterion of dissipated energy in the actuators of autonomous multi-legged robots used for service applications.

Others, like, Zhou et al. [201] proposed a new method called Friction Constraint Method (FriCoM) to find the optimal solution of feet forces for the grounded legs of a quadruped robot during locomotion. The method is efficient and computes solution at a faster rate compared to pseudo-inverse or incremental method. So, it is more suitable compared to pseudo-inverse or incremental method for real-time implementation. However, the method is not appreciable, since it did not consider the performance objective function. Further, Jiang et al. [79] used an exhaustive search method to obtain optimal feet forces in six-legged robots.

A few other researchers like Lapshin [100] proposed an energy efficient model of a walking machine and obtained few results from the point of view of gait-parameters only. Marhefka and Orin [114] selected optimal walking parameters for a six-legged robot on the basis of minimization of power over an entire locomotion cycle. They performed an analysis of average power consumption with respect to the parameters of wave gaits, based on a simulation model of the hexapod. Nishii [122] calculated optimal gait pattern based on minimization of the energetic cost of locomotion of legged robots. Here, the integral of weighted sum of the product of instantaneous joint torques and joint angular velocities were taken in addition to squares of the joint torques as the energetic cost, which was analyzed with respect to duty factor and velocity in the walking of a two joint six-legged robot. Also, Zhoga [200], Zelinski [189] and Silva et al. [159, 162] analyzed energy efficiency of multi-legged locomotion systems by taking leg dynamics and torque into account. However, in their studies, consideration of the type of joint actuator was not taken into account, although its contribution to energy consumption is significant. Lin and Song [109] developed the dynamic model of a quadruped walking robot and analysed the feet forces' distribution and energy efficiency with respect to duty factor and velocity. The robot's legs were massive, which meant effect of inertia was high and subsequently energy dissipation was also high.

Erden and Leblebicioglu [38] performed an energy efficiency analysis of a six-legged robot with wave gaits. Minimum energy consumption analysis for some specific gait strategies were carried out to determine the controlling parameters. Moreover, the authors proposed a phase modified wave gait for more energy efficient walking of six-legged robot. In the recent past, Santos et al. [146] developed an algorithm for six-legged robots moving on irregular terrain with alternating tripod gait and formulated the objective function as minimization of the energy consumption by using Nelder–Mead simplex method [97]. Hong and Cipra [67, 68] developed an analytical method to solve the multi-contact force distribution problem. Boyd and Wegbreit [22] proposed a new interior–point algorithm for contact force distribution. It has a complexity that is linear with the number of contact forces.

Though the above approaches give a vivid outlook of the foot force distribution and energy consumption analysis of legged robots, they possess certain limitations, which are listed below.

- Most of the studies on walking dynamics were conducted with simplified models of the legs and body. However, for a better understanding of the dynamics of the system, coupled-dynamical model of the robot that can study the inertia effects of a swing leg on the support legs and other swing legs is necessary to build.
- Kinematics and dynamic models based on a realistic foot-ground interaction of the robot in various terrain conditions are necessary to develop for addressing some important issues of walking, such as dynamic stability, energy efficiency and its on-line control.
- Validation of the developed model using VP tools prior to development of first physical prototype is yet to be done.
- Even though the attempts that were made by different researchers (as evident from the literature survey) were able to compute the optimal distribution of feet forces, they might not suitable for real-time implementations because the used optimization techniques were iterative in nature.

Therefore, there is a scope for improvement of the kinematic and dynamic models for estimation of optimal feet forces and joint torques along with energy efficiency, dynamic stability and on-line control of legged robots, vis-à-vis, six-egged robots.

On various terrains, locomotion only with straight-forward gait is not possible. So, omni-directional walking gaits like crab and turning must be used for accomplishing the task. Literature survey reveals that there are a very few studies on crab and turning gaits, though they are equally important for omni-directional maneuverability. In this regard, a few important studies like that of Hirose [63], Kumar and Waldron [95], Lee and Orin [104], Yoneda et al. [188], Saranli et al. [150], Kamikawa et al. [83], Estremera and Santos [39], Roy and Pratihar [142] are worth mentioning. However, to the best of author's knowledge, the studies on kinematic and dynamic analysis of crab and turning gaits of legged robots were not complete and confined to locomotion on flat terrain and that too, with simplified models. An attempt has been made in the present study to carry out kinematic and dynamic analysis of a six-legged robot with all these gaits on various terrains.

2.3 Stability Analysis of Multi-Legged Robots

Besides energy efficiency in multi-legged robots, maintaining stability is an important consideration in the development of any control algorithm for the multi-legged robot locomotion. As discussed above, agility is very much desirable in negotiating rough terrain. However, stability of the multi-legged system is also an obvious goal to achieve simultaneously, since toppling down carries risk of physical damage to the system along with slowing down to achieve the target. Six-legged robotic system provides better stability compared to other multi-legged robots (two-, three-, four-

or five-) and has a number of stable gait configurations. Several stability measures based on both static and dynamic stability have been carried by various researchers in the history of legged robots.

The stability of legged robots was first studied by McGhee and Frank [116] who defined the concept of static stability of an ideal walking robot on flat terrain using center of gravity projection method. The authors defined an ideal robot, which was assumed to have massless legs and no effect of system dynamics. According to center of gravity (COG) projection method, a vehicle is statically stable, if the projection of its COG lies within the support polygon. The method was extended to maneuverability over uneven terrain by Iswandhi and McGhee [77], where support polygon was defined by the horizontal projection of the real footholds. Thereafter, the concept of static stability margin (SSM) was introduced, which is the smallest perpendicular distance from the COG to the edge of the support polygon. However, SSM involves complex calculations, which Lee et al. [102] tried to avoid by approximating SSM to obtain longitudinal stability margin (LSM) for six-legged robots during straight-forward and crab motion. Similarly, Zhang and Song [190] adopted analytical and graphical methods to calculate LSM for wave-crab gait of a quadruped walking robot. LSM is nothing but the minimum distance from COG to the front and rear edges of the polygon taken along longitudinal axis. Min et al. [119] suggested a unique crab gait, which minimizes longitudinal gait stability margin according to variation in crab angle of a quadruped. The study also analyzed the effects of variations of footholds on stability and maximum permissible stroke. Further, Zhang and Song [191] introduced the concept of crab longitudinal stability margin (CLSM) giving a more logical reasoning by considering the walking robot as non-ideal, where inertia effects are significant. CLSM is the minimum distance from COG to the front and rear edges of the polygon taken along the direction of motion of the robot. Figure 2.1

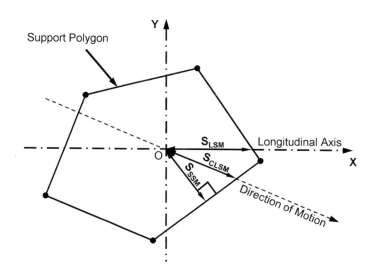

Fig. 2.1 Support polygon and static stability margins

shows support polygon and different static stability margin, namely SSM, LSM and CLSM.

All the above mentioned stability margins (SSM, LSM, CLSM etc.) were developed based on geometrical concepts and were independent of height of the COG and kinematic and dynamic parameters. However, for better stability measures, these parameters should not be ignored. A better stability measure called energy stability margin (ESM) was proposed by Messuri and Klein [117], which is defined as the minimum potential energy required by the robot to tip over the edges of the support polygon. Thereafter, Hirose et al. [66] generalized the stability criteria for walking robots by normalizing ESM using the robot's weight, which is commonly defined as normalized energy stability margin (NESM). Likewise, Roy and Pratihar [141, 142] analysed the stability of six-legged robots using the concept of NESM. A few other studies carried out, based on static stability for locomotion planning with crab and turning gaits are discussed, in brief.

Considering kinematic limit of joint motion and optimal gait stability margin, Lee and Bien [105] proposed a three-stage hierarchical strategy for planning crab gaits of a quadruped robot. From stability point of view, Santos and Jimenez [144] reported some advantages of crab and turning discontinuous gaits over continuous wave gaits of a quadruped robot. A new gait generation algorithm was proposed by Estremera et al. [40] for the development of free crab gaits for the hexapod robots on natural terrains. Some heuristic rules capable of planning leg motions accurately and guaranteeing stability for a non-periodic gait was included in conventional tripod gait to obtain a real free gait. These heuristic rules did rely on the definition of foot lifting instants and computation of new footholds.

In multi-legged robots, turning gaits are one of the most essential gaits for omni-directional walking. Orin [125] analyzed turning motion of a six-legged robot taking into consideration the crab angle. Hirose et al. [65] studied the turning motion of a quadruped robot. To optimize the turning gait, two criteria had been adopted in that study- (*a*) *Criteria 1: Achieving maximum rotational speed of the trunk body.* It had been assumed that, at any time only one leg of the robot was lifted and the returning speeds of all legs were equal though the returning paths were different in length. This led to different duty factor for different leg. Hence, to select the optimum duty factor for each leg, a nonlinear programming technique was used. (*b*) *Criteria 2: Maximizing the rotational angle of the trunk body in one cycle.* Since the angle of rotation was determined by the static stability and leg workspaces, a set of nonlinear equations and inequalities were established, and the variables were optimized using nonlinear programming techniques. The effectiveness of the technique had been proven through computer simulations and experiments using a quadruped walking vehicle model. Zhang and Song [190] studied the turning gait of a quadrupedal walking chair. The maximum rotation angle was determined by the leg workspaces alone, which was different from the approach used by Hirose et al. [65]. In this study, they adopted a combined analytical and graphical methods and developed several turning gaits. Zhang and Song [192] extended this work for stability analysis of spinning gaits and circling gaits of a quadruped walking machine. Cho et al. [31] proposed an optimal turning gait of a four-legged robot, which could minimize walking speed keeping stability margin larger than a certain required value. The proposed gait was sufficiently flexible, so as to permit crab walking, turning and pure rotational walking about a robot's geometric center. Miao and Howard [118] described an algorithm for generating a tripod turning gait, which could maximize the rotation angle and also optimize the stability. Recently, Yang [184, 185] studied crab gaits and turning gaits of a hexapod robot with a locked joint failure during tripod walking. In the recent past, a different static stability measure was carried out by Wang et al. [180] for locomotion of symmetric six-legged robots. A comparative study was carried out by the researchers with those of rectangular six-legged robots from the stability point of view. The authors defined the stability margin based on geometry for insect, mammalian and mixed gaits, as the longest step the robot can stretch without losing stability. Further, some studies used body position and posture control method to maintain the stability of the legged robots, as depicted in the studies of [25, 47, 58].

The above stability criterias are suitable, when the robot's dynamics is negligible. However, when a robot executes any kind of motion, the stability cannot be adjudged properly with static stability measures due to dynamic effects that arise due to its movements. Only dynamic stability criteria can give the better measures of its stability margin commonly named as dynamic stability margin (DSM), though efforts in this regard are meager. Vukobratovic and Juricic [175] were the first to propose dynamic stability criteria for biped robots based on zero-moment point (ZMP). Orin [126] proposed dynamic stability criteria for multi-legged robots, which is an extension of COG projection method. Momentum-based stability criteria had also been defined by a few researchers. Lin and Song [109] developed a dynamic model of a quadruped walking machine and tried to study the dynamic stability and energy efficiency for straight-forward motion using the concept of DSM. The dynamic stability margin showed a decreasing trend with the increasing velocity and decreasing stroke length. Koo and Yoon [92] also developed a dynamic model of a quadruped walking robot and tried to measure the stability for the dynamic gait at every instant using angular momentum principle, naming it as dynamic gait stability measure (DGSM). This work has been extended to measure the dynamic stability of hexapod walking robots in the present study. A few other dynamic stability criteria like Force-Angle Stability Margin (FASM) for mobile manipulators [128], Normalized Dynamic Energy Stability Margin (NDESM) for quadruped robots [48] etc. were also proposed that accounted for the presence of inertial effects in the system.

Hence, the aims and objectives of the present work are set as follows:

(a) Kinematic analysis including foot trajectory and gait planning of a realistic six-legged robot walking on various terrains like sloping surface, staircase and some user defined rough terrains etc.,

(b) Coupled non-linear dynamical modeling of robots with realistic three dimensional feet-ground contact model, considering visco-elastic compliance effect and slippage of feet tip,

(c) Studying the effects of walking parameters on energy consumption and static or dynamic stability of the said robot during gait generation over various terrains,

(d) Simulation of the different gait or walking patterns over various terrains using
     VP techniques and validation through real experiments.

     However, the present work does not focus on the following issues:

1. Frictions on the joints of the legs,
2. Large deformation and large leg-tip contact area during impact of feet tip with
    the ground,
3. Rebound on impact,
4. Compliance of entire leg of the robot.

2.4 Summary

This chapter provides an insight into the various studies carried out by researchers
worldwide on kinematics, dynamics, energy efficiency and stability of legged robots
moving on a variety of terrains with different walking gaits. The drawbacks have
clearly been brought out after thoroughly conducting a literature survey. This is
followed by both aims and objectives of the present work as well as an overview of
the subsequent chapters.

References

1. M. Ahmadi, M. Buehler, Stable control of a simulated one-legged running robot with hip and
    leg. IEEE Trans. Robot. Autom. **13**(1), 96–104 (1997)
2. F. Al-Bender, V. Lampaert, J. Swevers, The generalized Maxwell-slip model: a novel model
    for friction simulation and compensation. IEEE Trans. Autom. Control **50**(11), 1883–1887
    (2005)
3. R.M. Alexander, Three uses of springs in legged locomotion. Int. J. Robot. Res. **9**, 53–61
    (1990)
4. J.G. Alvarado, A.R. Agundis, H.R. Garduño, B.A. Ramírez, Kinematics of an asymmetrical
    three-legged parallel manipulator by means of the screw theory. Mech. Mach. Theory **45**,
    1013–1023 (2010)
5. M.Y. Al-Zaydi, S.H.M. Amin, *Simulation kinematics model of a multi-legged mobile robot*
    (IEEE Int. Conf. Adv. Robot., Monterey, CA, 1997), pp. 89–94
6. K. Arikawa, S. Hirose, Study of walking robot for 3 dimensional terrain, in *Proceedings of
    IEEE ICRA-95* (Nagoya, Japan, 1995), pp. 703–708
7. B. Armstrong-He´louvry, P. Dupont, C.C. de Wit, A survey of models, analysis tools and
    compensation methods for the control of machines with friction. Automatica **30**(7), 1083–
    1138 (1994)
8. J. Awrejcewicz, D. Grzelczyk, Y. Pyryev, A novel friction modelling and its impact on differ-
    ential equations computation and Lyapunov exponents estimation. J. Vibroengineering **10**(4),
    475–482 (2008)
9. M. Azad, R. Featherstone, A new nonlinear model of contact normal force. IEEE Trans.
    Robot. **30**(3), 736–739 (2014)
10. D.S. Bae, E.J. Haug, A recursive formulation for constrained mechanical system dynamics:
     part I open loop systems. Mech. Struct. Mach. **15**(3), 359–382 (1987)

11. J.P. Barreto, A. Trigo, P. Menezes, J. Dias, A.T. de Almeida, FBD-the free body diagram method. Kinematic and dynamic modeling of a six leg robot, in *Fifth International Workshop on Advanced Motion Control (AMC'98)* (Coimbra, Portugal, 1998), pp. 423–428
12. M. Bennani, F. Giri, Dynamic modelling of a four-legged robot. J. Intell. Rob. Syst. **17**(4), 419–428 (1996)
13. C.A. Berardi-Gonziilez, H. Martinez-Alfaro, Kinematic simulator for an insect-like robot, in *IEEE International Conference on Systems, Man and Cybernetics,* vol. 2 (Washington DC, USA, 2003), pp. 1846–1851
14. P. Bergés, A. Bowling, Rebound, slip, and compliance in the modeling and analysis of discrete impacts in legged. J. Vib. Control **12**(12), 1407–1430 (2006)
15. M.D. Berkemeier, Modeling the dynamics of quadrupedal running. Int. J. Robot. Res. **17**, 971–985 (1998)
16. S.-S. Bi, X.-D. Zhou, D.B. Marghitu, Impact modelling and analysis of the compliant legged robots, in *Proceedings of the Institution of Mechanical Engineers, Part K: Journal of Multibody Dynamics*, **226**(2):85–94 (2012)
17. P.T. Bibalan, R. Featherstone, A study of soft contact models in Simulink, in *Australasian Conference on Robotics and Automation (ACRA)* (Sydney, Australia, 2009), p. 8
18. M. Boegli, T. de Laet, J. de Schutter, J. Swevers, A smoothed GMS friction model for moving horizon friction state and parameter estimation, in *The 12th IEEE International Workshop on Advanced Motion Control* (Sarajevo, Bosnia and Hercegovina, 2012)
19. Q. Bombled, O. Verlinden, Dynamic simulation of six-legged robots with a focus on joint friction. Multibody Sys. Dyn. **28**(4), 395–417 (2006)
20. A.P. Bowling, Dynamic performance, mobility, and agility of multilegged robots. ASME J. Dyn. Syst., Meas., Control. **128**, 765–777 (2006)
21. A.P. Bowling, *Impact forces and mobility in legged robot locomotion* (IEEE/ASME Int. Conf. Adv. Intell. Mechatron., Zurich, Switzerland, 2007), pp. 1–8
22. S.P. Boyd, B. Wegbreit, Fast computation of optimal contact forces. IEEE Trans. Rob. **23**(6), 1117–1132 (2007)
23. R.M. Brach, Mechanical impact dynamics: rigid body collisions: Brach Engineering (LLC, 2007)
24. W.C. de Canudas, H. Olsson, K. Aström, P. Lischinsky, A new model for control of systems with friction. IEEE Trans. Autom. Control **40**(5), 419–425 (1995)
25. G. Chen, B. Jin, Y. Chen, Accurate position and posture control of a redundant hexapod robot. Arab. J. Sci. Eng. **42**(5), 2031–2042 (2017)
26. J.-S. Chen, F.-T. Cheng, K.-T. Yang, F.-C. Kung, S. York-Yih, Optimal force distribution in multilegged vehicles. Robotica **17**(2), 159–172 (1999)
27. X. Chen, K. Watanabe, K. Kiguchi, K. Izumi, optimal force distribution for the legs of a quadruped robot. Mach. Intell. Robot. Control. **1**(2), 87–94 (1999)
28. F.T. Cheng, D.E. Orin, Optimal force distribution in multiple chain robotic systems. IEEE Trans. Syst. Man Cybern. **21**(1), 13–24 (1991)
29. F.-T. Cheng, T.-H. Chen, Y.-Y. Sun, Resolving manipulator redundancy under inequality constraints. IEEE Trans. Robot. Autom. **10**(1), 65–71 (1994)
30. Z. Chi, D. Zhang, L. Xia, Z. Gao, Multi-objective optimization of stiffness and workspace for a parallel kinematic machine. Int. J. Mech. Mater. Des. **9**(3), 281–293 (2013)
31. D.J. Cho, J.H. Kim, D.G. Gweon, Optimal turning gait of a quadruped walking robot. Robotica **13**(6), 559–564 (1995)
32. P. Dahl, *A solid friction model, Technical Report TOR-0158H3107–18I-1* (The Aerospace Corporation, El Segundo, CA, 1968)
33. J. Denavit, R.S. Hartenberg, A kinematic notation for lower-pair mechanisms based on matrices. Trans. ASME J Appl. Mech. **23**, 215–221 (1955)
34. L. Ding, H. Gao, Z. Deng, J. Song, Y. Liu, G. Liu, K. Iagnemma, Foot–terrain interaction mechanics for legged robots: modeling and experimental validation. Int. J. Robot. Res. **32**(13), 1585–1606 (2013)

35. S. Djerassi, Three-dimensional, one point collision with friction. Multibody Sys. Dyn. **27**(2), 173–195 (2012)
36. P. Dupont, V. Hayward, B. Armstrong, F. Altpeter, Single state elasto-plasticfric tion models. IEEE Trans. Autom. Control **47**(5), 787–792 (2002)
37. M.S. Erden, K. Leblebicioglu, Torque distribution in a six-legged robot. IEEE Trans. Rob. **23**(1), 179–186 (2007)
38. M.S. Erden, K. Leblebicioglu, Analysis of wave gaits for energy efficiency. Auton. Robot. **23**, 213–230 (2007)
39. J. Estremera, P.G. de Santos, Generating continuous free crab gaits for quadruped robots on irregular terrain. IEEE Trans. Rob. **21**(6), 1067–1076 (2005)
40. J. Estremera, J.A. Cobano, P.G. de Santos, Continuous free-crab gaits for hexapod robots on a natural terrain with forbidden zones: An application to humanitarian demining. Robot. Auton. Syst. **58**(5), 700–711 (2010)
41. R. Featherstone, *Rigid Body Dynamics Algorithms* (Springer, New York, 2008)
42. R. Featherstone, *Robot Dynamics Algorithms*, (Kluwer Academic Publishers, Boston/ Dordrecht/ Lancaster, 1987)
43. G. Figliolini, S.D. Stan, P. Rea, Motion analysis of the leg tip of a six-legged walking robot, in *12th IFToMM World Congress* (Besançon, France, 2007)
44. P.S. Freeman, D.E. Orin, Efficient dynamic simulation of a quadruped using a decoupled tree structured approach. Int. J. Robot. Res. **10**, 619–627 (1991)
45. R.J. Full, M.S. Tu, Mechanics of a rapid running insect: two-, four and sex-legged locomotion. J. Exp. Biol. **156**, 215–231 (1991)
46. J.A. Galvez, J. Estremera, P.G. de Santos, A new legged-robot configuration for research in force distribution. Mechatronics **13**, 907–932 (2003)
47. C. Gang, J. Bo, C. Ying, Nonsingular fast terminal sliding mode posture control for six-legged walking robots with redundant actuation. Mechatronics **50**, 1–15 (2018)
48. E. Garcia, P.G. de Santos, An improved energy stability margin for walking machines subject to dynamic effects. Robotica **23**(1), 13–20 (2005)
49. M.C. García-López, E. Gorrostieta-Hurtado, E. Vargas-Soto, J.M. Ramos-Arreguín, A. Sotomayor-Olmedo, J.C. Moya-Morales, Kinematic analysis for trajectory generation in one leg of a hexapod robot. Procedia Technol. **3**, 342–350 (2012)
50. J. Gardner, Efficient computation of force distributions for walking machines on rough terrain. Robotica **10**(5), 427–433 (1992)
51. J.F. Gardner, Force distribution in walking machines over rough terrain. ASME J. Dyn. Syst. Meas. Control **113**(4), 754–758 (1991)
52. L. Gaul, M. Mayer, Efficient modelling of contact interfaces of joints in built-up structures. in *Computer Methods and Experimental Measurements for surface effects and contact mechanics VIII*, ed. J.T.M. de Hosson, C.A. Brebbia, S.-I. Nishida (WIT Press, Southampton, Boston, 2007), pp. 195–205
53. G. Gilardi, I. Sharf, Literature survey of contact dynamics modelling. Mech. Mach. Theory **37**(10), 1213–1239 (2002)
54. W. Goldsmith, *Impact: The Theory and Physical Behaviour of Colliding Solids* (Dover Publications Inc, Mineola, New York, 2001)
55. Y. Gonthier, J. Mcphee, C. Lange, J.-C. Piedboeuf, A regularized contact model with asymmetric damping and dwell-time dependent friction. Multibody Sys. Dyn. **11**(3), 209–233 (2004)
56. M.M. Gor, P.M. Pathak, A.K. Samantaray, J.-M. Yang, S.W. Kwak, Control oriented model-based simulation and experimental studies on a compliant legged quadruped robot. Robot. Autom. Syst. **72**:217–234 (2015)
57. D.M. Gorinevsky, A.Y. Shneider, Force control in locomotion of legged vehicles over rigid and soft surfaces. Int. J. Robot. Res. **9**(2), 4–23 (1990)
58. C. Grand, F. Benamar, F. Plumet, P. Bidaud, Stability and traction optimization of a reconfigurable wheel-legged robot. Int. J. Robot. Res. **23**(10–11), 1041–1058 (2004)

59. S.R. Hamner, A. Seth, K.M. Steele, S.L. Delp, A rolling constraint reproduces ground reaction forces and moments in dynamic simulations of walking, running, and crouch gait. J. Biomech. **46**(10):1772–1776 (2013)
60. I. Han, B.J. Gilmore, Multi-body impact motion with friction—analysis, simulation and experimental validation. ASME J. Mech. Des. **115**(3), 412–422 (1993)
61. J.P.D. Hartog, Forced vibrations with combined Coulomb and viscous friction. Trans. ASME Appl. Mech **53**, 107–115 (1931)
62. K. Hauser, T. Bretl, J.C. Latombe, K. Harada, B. Wilcox, Motion planning for legged robots on varied terrain. Int. J. Robot. Res. **27**(11–12), 1325–1349 (2008)
63. S. Hirose, A study of design and control of a quadruped walking vehicle. Int. J. Robot. Res. **3**(2), 113–132 (1984)
64. S. Hirose, Y. Umetani, Some consideration on a feasible walking mechanism as a terrain vehicle, in *Proceedings of 3rd International CISM-IFToMM Symposium* (Udine, Italy, 1978), pp. 357–375
65. S. Hirose, H. Kikuchi, Y. Umetani, The standard circular gait of a quadruped walking vehicle. Adv. Robot. **1**(2), 143–164 (1986)
66. S. Hirose, H. Tsukagoshi, K. Yoneda, Normalized energy stability margin: generalized stability criterion for walking vehicles, in *Proceedings of the International Conference on Climbing and Walking Robots* (Brussels, Belgium, 1998), pp. 71–76
67. D.W. Hong, R.J. Cipra, Visualization of the contact force solution space for multi-limbed robots. ASME J. Mech. Des. **128**(1), 295–302 (2006)
68. D.W. Hong, R.J. Cipra, Optimal contact force distribution for multi-limbed robots. ASME J. Mech. Des. **128**(3), 566–573 (2006)
69. D. Howard, S.J. Zhang, D.J. Sanger, Kinematic analysis of a walking machine. Math. Comput. Simul. **41**(5), 525–538 (1996)
70. W. Hu, D.W. Marhefka, D.E. Orin, Hybrid kinematic and dynamic simulation of running machines. IEEE Trans. Rob. **21**(3), 490–497 (2005)
71. Z. Huang, Y.S. Zhao, The accordance and optimization-distribution equations of the over-determinate inputs of walking machines. Mech. Mach. Theory **29**(2), 327–332 (1994)
72. M.H. Hung, D.E. Orin, K.J. Waldron, Efficient formulation of the force distribution equations for general tree-structured robotic mechanisms with a mobile base. IEEE Trans. Syst. Man Cybern.-Part B. **30**(4), 529–538 (2000)
73. K. Hunt, F. Crossley, Coefficient of restitution interpreted as damping in vibroimpact. Trans. ASME-J. Appl. Mech. **42**, 440–445 (1975)
74. R.L. Huston, *Multibody Dynamics* [Book]. (Butterworth–Heinemann, Boston, MA, 1990)
75. M.L. Husty, An algorithm for solving the direct kinematics of general Stewart-Gough platforms. Mech. Mach. Theory **31**(4), 365–379 (1996)
76. C. Innocenti, V. Parenti-Castelli, Direct position analysis of the Stewart platform mechanism. Mech. Mach. Theory **25**(6), 611–621 (1990)
77. R. Iswandhi, G. McGhee, Adaptive locomotion for a multilegged robot over rough terrain. IEEE Trans. Syst. Man Cybern. **9**(4), 176–182 (1979)
78. W.D. Iwan, A distributed-element model for hysteresis and its steady-state dynamic response. J. Appl. Mech. **33**, 893–900 (1966)
79. A. Jiang, M. Liu, D. Howard, Optimization of legged robot locomotion by control of foot-force distribution. Trans. Inst. W. Y. Meas. Control. **26**(4), 311–323 (2004)
80. B. Jin, C. Chen, W. Li, Power consumption optimization for a hexapod walking robot. J. Intell. Robot. Syst. **71**(2), 195–209 (2013)
81. L. Jingtao, W. Feng, Y. Huangying, W. Tianmiao, Y. Peijiang, Energy efficiency analysis of quadruped robot with trot gait and combined cycloid foot trajectory. Chin. J. Mech. Eng. **27**(1), 138–145 (2014)
82. M. Kalakrishnan, J. Buchli, P. Pastor, M. Mistry, S. Schaal, Learning, planning, and control for quadruped locomotion over challenging terrain. Int. J. Robot. Res. 30(2), 236–258 (2011)
83. K. Kamikawa, T. Arai, K. Inoue, Y. Mae, Omni-directional gait of multi-legged rescue robot, in *Proceedings of IEEE International Conference on Robotics and Automation* (New Orleans, LA, 2004), pp. 2171–2176

84. T.R. Kane, C.F. Wang Source, On the derivation of equations of motionJ. Soc. Ind. Appl. Math. **13**(2), 487–492 (1965)
85. D.C. Kar, K.K. Issac, K. Jayarajan, Gaits and energetics in terrestrial legged locomotion. Mech. Mach. Theory **38**(2), 355–366 (2003)
86. D.C. Kar, K.K. Issac, K. Jayarajan, Minimum energy force distribution for a walking robot. J. Robot. Syst. **18**(2), 47–54 (2001)
87. R. Kelaiaia, O. Company, A. Zaatri, Multiobjective optimization of a linear delta parallel robot. Mech. Mach. Theory **50**, 159–178 (2012)
88. J. Kerr, B. Roth, Analysis of multifingered hands. Int. J. Robot. Res. **4**(4), 3–17 (1986)
89. R. Kikuuwe, N. Takesue, A. Sano, H. Mochiyama, H. Fujimoto, Fixed-step friction simulation: from classical coulomb model to modern continuous models, in *Proceeding of IEEE/RSJ International Conference on Intelligent Robots and Systems* (Edmonton, Alta., Canada, 2005), pp. 1009–1016
90. S.W. Kim, Contact dynamics and force control of flexible multi-body systems, Ph.D. thesis, Department of Mechanical Engineering, McGill University, Montreal 1999
91. C.A. Klein, S. Kittivatcharapong, Optimal force distribution for the legs of a walking machine with friction cone constraints. IEEE Trans. Robot. Autom. **6**(1), 73–85 (1990)
92. T.-W. Koo, Y.-S. Yoon, Dynamic instant gait stability measure for quadruped walking robot. Robotica **17**, 59–70 (1999)
93. P.R. Kraus, A. Fredriksson, V. Kumar, Modeling of frictional contacts for dynamic simulation, in *Proceedings of IROS 1997 Workshop on Dynamic Simulation: Methods and Applications* (1997)
94. V. Kumar, K.J. Waldron, Force distribution in walking vehicles. ASME J. Mech. Des. **112**, 90–99 (1990)
95. V. Kumar, K.J. Waldron, Gait analysis for walking machines for omnidirectional locomotion on uneven terrain, in *7th CISM-IFToMM Symposium on Theory and Practice of Robots and Manipulators* (Udine, Italy, 1988), pp. 37–62
96. R. Kurazume, K. Yoneda, S. Hirose, Feedforward and feedback dynamic trot gait control for quadruped walking vehicle. Auton. Robot. **12**(2), 157–172 (2002)
97. J.C. Lagarias, J.A. Reeds, M.H. Wright, P.E. Wright, Convergence properties of the nelder-mead simplex method in low dimensions. SIAM J. Optim. **9**(1), 112–147 (1998)
98. V. Lampaert, J. Swevers, F. Al-Bender, Modification of the Leuven integrated friction model structure. IEEE Trans. Autom. Control **47**(4), 683–687 (2002)
99. H.M. Lankarani, P.E. Nikravesh, Continuous contact force models for impact analysis in multi-body systems. Nonlinear Dyn. **5**, 193–207 (1994)
100. V.V. Lapshin, Energy consumption of a walking machine: model estimations and optimization, in *Proceedings of 7th Conference on Advanced Robotics* (San Feliu de Guixols, 1995), pp. 420–425
101. J.K. Lee, S.M. Song, A study of instantaneous kinematics of walking machines. Int. J. Robot. Autom. **5**(3), 131–138 (1990)
102. T.T. Lee, C.M. Liao, T.K. Chen, On the stability properties of hexapod tripod gait. IEEE J. Robot. Autom. **4**(4), 427–434 (1988)
103. T.W. Lee, A.C. Wang, On the dynamics of intermittent-motion mechanisms—part 1: dynamic model and response. J. Mech. Transmiss. Autom. Des. **105**(3), 534–540 (1983)
104. W.J. Lee, D.E. Orin, Omni-directional supervisory control of a multi-legged vehicle using periodic. IEEE Int. J. Robot. Autom. **4**(6), 635–642 (1988)
105. Y.J. Lee, Z. Bien, A hierarchical strategy for planning crab gaits of a quadruped walking robot. Robotica **12**, 23–31 (1994)
106. H. Lehtinen, Force based motion control of a walking machine, Finland, ESPOO: Ph.D. thesis, (Technical Research Centre, 1994)
107. K. Li, X. Ding, M. Ceccarell, A total torque index for dynamic performance evaluation of a radial symmetric six-legged robot. Front. Mech. Eng. **7**(2), 219–230 (2012)
108. Z. Li, S.S. Ge, S. Liu, Contact-force distribution optimization and control for quadruped robots using both gradient and adaptive neural networks, in IEEE Trans. Neural Netw. Learn. Syst. **25**(8), 1460–1473 (2014)

109. B.-S. Lin, S.-M. Song, Dynamic modeling, stability and energy efficiency of a quadrupedal walking machine. J. Robotic Syst. **18**(11), 657–670 (2001)
110. H. Liu, B. Wen, Force distribution for the legs of a quadruped walking vehicle. J. Robot. Syst. **14**(1), 1–8 (1997)
111. V. Loc, S. Roh, I.M. Koo, D.T. Tran, H.M. Kim, H. Moon, H.R. Choi, Sensing and gait planning of quadruped walking and climbing robot for traversing in complex environment. Robot. Auton. Syst. **58**(5), 666–675 (2010)
112. C. Mahfoudi, K. Djouani, S. Rechak, M. Bouaziz, Optimal force distribution for the legs of an hexapod robot, in *Proceedings of IEEE Conference on Control Applications* (Istanbul, Turkey, 2003), pp. 657–663
113. D.W. Marhefka, D.E. Orin, A compliant contact model with nonlinear damping for simulation of robotic systems. IEEE Trans. Syst. Man Cybern. Part A Syst. Hum. **29**, 566–572 (1999)
114. D.W. Marhefka, D.E. Orin, Gait planning for energy efficiency in walking machines, in *Proceedings of IEEE International Conference on Robotics and Automation* (Albuquerque, New Mexico, 1997), pp. 474–480
115. D.W. Marhefka, D.E. Orin, Quadratic optimization of force distribution in walking machines, in *Proceedings of IEEE International Conference on Robotics and Automation* (Lueven, Belgium, 1998), pp. 477–483
116. R.B. McGhee, A.A. Frank, On the stability properties of quadruped creeping gaits. Math. Biosci. **3**, 331–351 (1968)
117. D. Messuri, C. Klein, Automatic body regulation for maintaining stability of a legged vehicle during rough-terrain locomotion. *IEEE J. Robot. Autom.* RA-**1**(3), 132–141 (1985)
118. S. Miao, D. Howard, Optimal tripod turning gait generation for hexapod walking machines. Robotica **18**(6), 639–649 (2000)
119. B. Min, Z. Bein, S. Hwangt, Basic characteristics and stability properties of quadruped crab gaits. Robotica **11**(3), 233–243 (1993)
120. M.A. Nahon, J. Angeles, Real-time force optimization in parallel kinematic chains under inequality constraints. IEEE Trans. Robot. Autom. **8**(4), 439–450 (1992)
121. H. Nijmeijer, A. Rodriguez-Angeles, Synchronization of Mechanical Systems, *World Scientific Series on Nonlinear Science,* vol. 46 (Singapore, Ed. Leon O. Chua, 2003) Series A
122. J. Nishii, Gait pattern and energetic cost in hexapods, in *Proceedings of 20th Annual International Conference of the IEEE Engineering in Medicine and Biology Society* (Hong Kong, China, 1998), pp. 2430–2433
123. H. Olsson, K.J. Åström, C.C. de Wit, M. Gäfvert, P. Lischinsky, Friction models and friction compensation. Eur. J. Control. **4**(3), 176–195 (1998)
124. D.E. Orin, Supervisory control of a multi-legged robot. Int. J. Robot. Res. **1**(4), 79–91 (1982)
125. D.E. Orin, Y. Oh, Control of force distribution in robotic mechanisms containing closed kinematic chains. ASME J. Dyn. Syst. Meas. Control. **103**(2), 134–141 (1981)
126. D. Orin, Interactive control of a six-legged vehicle with optimization of both stability and energy, Ph.D. thesis, The Ohio State University 1976
127. F.B. Ouezdou, O. Bruneau, J.C. Guinot, Dynamic analysis tool for legged robots. Multibody Sys. Dyn. **2**(4), 369–391 (1998)
128. E. Papadopoulos, D. Rey, A new measure of tipover stability margin for mobile manipulators, in *Proceedings of IEEE International Conference Robotics and Automation* (Minneapolis, Minnesota, 1996), pp. 3111–3116
129. F. Pfeiffer, H.-J. Weidemann, P. Danowski, Dynamics of the walking stick insect, in *Proceeding of the IEEE Internatiaonal Coriference on Robotics and Automation* (Cincinatti, Ohio, 1990), pp. 1458–1463
130. D.K. Pratihar, K. Deb, A. Ghosh, Optimal turning gait of a six-legged robot using GA-fuzzy approach. Artif. Intell. Eng. Des. Anal. Manuf. **14**(3), 207–219 (2000)
131. A. Preumont, P. Alexadre, D. Ghuys, Gait analysis and implementation of a six LegWalking machine, 91IICAR, in *Fifth International Conference on Advanced Robotics—Robots in Unstructured Environments* (Pisa, Italy, 1991), pp. 941–945

132. A. Preumont, P. Alexandre, I. Doroftei, F. Goffin, A conceptual walking vehicle for planetory exploration. Mechatronics **7**(3), 287–296 (1997)

133. S. Qian, B. Zi, D. Zhang, L. Zhang, Kinematics and error analysis of cooperative cable parallel manipulators for multiple mobile cranes. Int. J. Mech. Mater. Des. **10**(4), 395–409 (2014)

134. E.Y. Raby, D.E. Orin, Passive walking with leg compliance for energy-efficient multilegged vehicles, in *Proceedings of IEEE International Conference on Robotics and Automation* (Detroit, USA, 1999), pp. 1702–1707

135. M.H. Raibert, *Legged Robots That Balance*, 1st edn. (The MIT Press, Cambridge, 1986)

136. M. Rajaei, H. Ahmadian, Development of generalized Iwan model to simulate frictional contacts with variable normal loads. Appl. Math. Model. **38**(15–16), 4006–4018 (2014)

137. L. Rolland, Certified solving of the forward kinematics problem with an exact algebraic method for the general parallel manipulator. Adv. Robot. **19**(9), 995–1025 (2005)

138. D.E. Rosenthal, An order n formulation for robotic systems. J. Astronaut. Sci. **38**(4), 511–529 (1990)

139. T. Rossmann, F. Pfeiffer, C. Glocker, Efficient algorithms for non-smooth dynamics. In *Proceedings of the ASME International Mechanical Engineering Congress and Exposition* (Dallas, Texas, 1997)

140. S.S. Roy, D.K. Pratihar, Effects of turning gait parameters on energy consumption and stability of a six-legged walking robot. Robot. Auton. Syst. **60**(1), 72–82 (2012)

141. S.S. Roy, D.K. Pratihar, Dynamic modeling, stability and energy consumption analysis of a realistic six-legged walking robot. Robot. Comput. Integr. Manuf. **29**(2), 400–416 (2013)

142. S.S. Roy, D.K. Pratihar, Kinematics, dynamics and power consumption analyses for turning motion of a six-legged robot. J. Intell. Rob. Syst. **74**(3–4), 663–668 (2014)

143. S.S. Roy, A.K. Singh, D.K. Pratihar, Estimation of optimal feet forces and joint torques for on-line control of six-legged robot. Robot. Comput. Integr. Manuf. **27**(5), 910–917 (2011)

144. E. Sacks, L. Joskowicz, Computational kinematic analysis of higher pairs with multiple contacts. ASME J. Mech. Des. **117**(2A), 269–277 (1995)

145. X.Y. Sandoval-Castro, M. Garcia-Murillo, L.A. Perez-Resendiz, E. Castillo-Castañeda, Kinematics of hex-piderix-a six-legged robot-using screw theory. Int. J. Adv. Rob. Syst. **10**(19), 1–8 (2013)

146. P.G. de Santos, E. Garcia, R. Ponticelli, M. Armada, Minimizing energy consumption in hexapod robots. Adv. Robot. **23**(6), 681–704 (2009)

147. P.G. de Santos, M.A. Jimenez, Path tracking with quadruped walking machines using discontinuous gaits. Comput. Electr. Eng. **21**(6), 383–396 (1995)

148. P.G. de Santos, J. Estremera, E. Garcia, Optimizing leg distribution around the body in walking robots, in *Proceedings of IEEE International Conference on Robotics and Automation* (Barcelona, Spain, 2005), pp. 3207–3212

149. U. Saranli, M. Buehler, Modeling and analysis of a spatial compliant hexapod, in *Technical papers, Department* of Electrical *Engineering and Computer Science (*McGill University, Montreal, Canada, 1999), pp. 1–18

150. U. Saranli, M. Buehler, D.E. Koditschek, RHex: a simple and highly mobile hexapod robot. Int. J. Robot. Res. **20**(7), 616–631 (2001)

151. J.P. Schmiedeler, K.J. Waldron, Impact analysis as a design tool for the legs of mobile robots, in *Advances in Robot Kinematics* ed. by J. Lenarcic, M.M. Stanisic (Kluwer Academic Publishers, Norwell, MA, 2000)

152. D.J. Segalman, Modeling contact friction in structural dynamics. Struct. Control. Health Monit **13**(1), 430–453 (2006)

153. J.E. Shigley, R.G. Budynas, J.K. Nisbett, *Mechanical Engineering Design*, 9th edn. (McGraw-Hill, New York, 2011)

154. E. Shin, D.A. Streit, An energy of spring efficient quadruped with two stage equilibrator. ASME J. Mech. Des. **115**(1), 156–163 (1993)

155. A. Shkolnik, R. Tedrake, Inverse kinematics for point foot quadruped robot with dynamic redundancy resolution, in *IEEE International Conference Robotics and Automation, (Rome, Italy*, 2007), pp. 4331–4336

156. A. Shkolnik, M. Levashov, I.R. Manchester, R. Tedrake, Bounding on rough terrain with the LittleDog robot. Int. J. Robot. Res. **30**(2), 192–215 (2011)
157. M.F. Silva, J.A.T. Machado, A literature review on the optimization of legged robots. J. Vib. Control **18**(12), 1753–1767 (2012)
158. M.F. Silva, J.A.T. Machado, R.S. Barbosa, Complex order dynamics of hexapod locomotion. Sig. Process. **86**(10), 2785–2793 (2006)
159. M.F. Silva, J.A.T. Machado, A.M. Lopes, Energy analysis of multi-legged locomotion systems, in *Proceedings of 4*th *Conference on Climbing and Walking Robots* (Karlsruhe, Germany, 2001), pp. 143–150
160. M.F. Silva, J.A.T. Machado, A.M. Lopes, Fractional order control of a hexapod robot. Nonlinear Dyn. **38**(1–4), 417–433 (2004)
161. M.F. Silva, J.A.T. Machado, A.M. Lopes, Modelling and simulation of artificial locomotion systems. Robotica **23**(5), 595–606 (2005)
162. M.F. Silva, J.A.T. Machado, A.M. Lopes, Performance analysis of multi-legged systems, in *Proceedings of IEEE 2nd International Workshop on Robot Motion and Control* (Bukowy Dworek, Poland, 2001), pp. 45–50
163. M.F. Silva, J.A.T. Machado, A.M. Lopes, Position/Force Contro of a Walking_Robot. Machine Intelligence and Robotic Control **5**(2), 33–44 (2003)
164. P. Song, P. Kraus, V. Kumar, P. Dupont, Analysis of rigid-body dynamic models for simulation of systems with frictional contacts. ASME J. Appl. Mech. **68**(1), 118–128 (2001)
165. S.M. Song, V.J. Vohnout, K.J. Waldron, G.L. Kinzel, Computer-aided design of a leg for an energy efficient walking machine. Mech. Mach. Theory **19**(1), 17–24 (1994)
166. S. Soyguder, H. Alli, Kinematic and dynamic analysis of a hexapod walking–running–bounding gaits robot and control actions. Comput. Electr. Eng. **38**(2), 444–458 (2012)
167. W.J. Stronge, *Impact Mechanics* (Cambridge University Press, 2000)
168. W.J. Stronge, Smooth dynamics of oblique impact with friction. Int. J. Impact Engg. **51**, 36–49 (2013)
169. Y. Sugahara, G. Carbone, K. Hashimoto, M. Ceccarelli, H.O. Lim, A. Takanishi, Experimental stiffness measurement of WL-16RII biped walking vehicle during walking operation. J. Robot. Mechatron. **19**(3), 272–280 (2007)
170. J. Swevers, F. Al-Bender, C. Ganseman, T. Prajogo, An integrated friction model structure with improved presliding behavior for accurate friction compensation. IEEE Trans. Autom. Control **45**(4), 675–686 (2000)
171. M. Tarokh, M. Lee, Systematic method for kinematics modelling of legged robots on uneven terrain. Int. J. Control. Autom. **2**(2), 9–18 (2009)
172. J.C. Trinkle, J.S. Pang, S. Sudarsky, G. Lo, On dynamic multi-rigid-body contact problems with coulomb friction. Math. Mech. **77**(4), 267–279 (1997)
173. V. Vasilopoulos, I.S. Paraskevas, E.G. Papadopoulos, Compliant terrain legged locomotion using a viscoplastic approach, in *IEEE/RSJ International Conference on Intelligent Robots and Systems (IROS '14)* (Chicago, Illinois, USA, 2014), pp. 4849–4854
174. R. Vidoni, A. Gasparetto, Efficient force distribution and leg posture for a bio-inspired spider robot. Robot. Auton. Syst. **59**(2), 142–150 (2011)
175. M. Vukobratovic, D. Juricic, Contribution to the synthesis of biped gait, *IEEE Trans. Biomed. Eng.* **BME-16**(1), 1–6 (1969)
176. M. Vukobratovic, V. Potkonjak, K. Babkovic, B. Borovac, Simulation model of general human and humanoid motion. Multibody Sys. Dyn. **17**(1), 71–96 (2007)
177. K.J. Waldron, M. Raghavan, B. Roth, Kinematics of a hybrid series-parallel manipulation system. Trans. ASME J. Dyn. Syst. Meas. Control. **111**(2), 211–221 (1989)
178. M.W. Walker, D.E. Orin, Efficient dynamic computer simulation of robotic mechanisms. J. Dyn. Syst. Meas. Contr. **104**(3), 205–211 (1982)
179. H. Wang, L. Sang, X. Hu, D. Zhang, H. Yu, Kinematics and dynamics analysis of a quadruped walking robot with parallel leg mechanism. Chin. J. Mech. Eng. **26**(5), 881–891 (2013)
180. Z.Y. Wang, X.L. Ding, A. Rovetta, Analysis of typical locomotion of a symmetric hexapod robot. Robotica **28**(6), 893–907 (2010)

181. Z.Y. Wang, X.L. Ding, A. Rovetta, A. Giusti, Mobility analysis of the typical gait of a radial symmetrical six-legged robot. Mechatronics **21**(7), 1133–1146 (2011)
182. X. Xu, Y. Chen, A method for trajectory planning of robot manipulators in cartesian space, in *Proceedings of the 3d World Congress on Intelligent Control and Automation, June 28-July 2* (Hefei, P.R. China, 2000), pp. 1220–1225
183. B.D. Yang, M.L. Chu, C.H. Menq, Stick–slip–separation analysis and non-linear stiffness and damping characterization of friction contacts having variable normal load. J. Sound Vib. **210**, 461–481 (1998)
184. J.M. Yang, Fault-tolerant crab gaits and turning gaits for a hexapod robot. Robotica **24**, 269–270 (2006)
185. J.M. Yang, Omnidirectional walking of legged robots with a failed leg. Math. Comput. Model. **47**, 1372–1388 (2008)
186. A.S. Yigit, A.P. Christoforou, M.A. Majeed, A nonlinear visco-elastoplastic impact model and the coefficient of restitution. Nonlinear Dyn. **66**(4), 509–521 (2011)
187. K. Yoneda, H. Iiyama, S. Hirose, Intermittent trot gait of a quadruped walking machine dynamic stability control of an omnidirectional walk, in *Proceedings of the IEEE International Conference on Robotics and Automation* (Minneapolis, Minnesota, 1996), pp. 3002–3007
188. K. Yoneda, K. Suzuki, Y. Kanayama, H. Takanishi, J. Akizono, Gait and foot trajectory planning for versatile motions of a sixlegged robot. J. Robot. Syst. **14**(2), 121–133 (1997)
189. T. Zelinska, Efficiency analysis in the design of walking machine. J. Theor. Appl. Mech. **38**, 693–708 (2000)
190. C. Zhang, S. Song, Gaits and geometry of a walking chair for the disabled. J. Terrramech. **26**(3–4), 211–233 (1989)
191. C.D. Zhang, S.M. Song, Stability analysis of wave-crab gaits of a quadruped. J. Robotic Syst. **7**(2), 243–276 (1990)
192. C.D. Zhang, S.M. Song, Turning gait of a quadrupedal walking machine, in *Proceedings of IEEE International Conference on Robotics and Automation* (Sacramento, California, 1991), pp. 2106–2112
193. D. Zhang, Z. Gao, Forward kinematics, performance analysis and multi-objective optimization of a bio-inspired parallel manipulator. Robot. Comput.-Integr. Manuf. **28**, 484–492 (2012)
194. D. Zhang, Z. Gao, I. Fassi, Design optimization of a spatial hybrid mechanism for micromanipulation. Int. J. Mech. Mater. Des. **7**(1), 55–70 (2011)
195. D. Zhang, X.M. Su, Z. Gao, J.J. Qian, Design, analysis and fabrication of a novel three degrees of freedom parallel robotic manipulator with decoupled motions. Int. J. Mech. Mater. Des. **9**(3), 199–212 (2013)
196. S.J. Zhang, D. Howard, D.J. Sanger, D.R. Kerr, S. Miao, Walking machine design based on the mechanics of the Stewart platform, in *Proceedings of ASME ESDA Conference* (London, 1994), pp. 849–855
197. Y.N. Zhang, I. Sharf, Validation of nonlinear viscoelastic contact force models for low speed impact. ASME J. Appl. Mech. **76**(5), 051002–1-051002-12 (2009)
198. Y.S. Zhao, Lu Ling, T.S. Zhao, Y.H. Du, Z. Huang, Dynamic performance analysis of six-legged walking machines. Mech. Mach. Theory **35**(1), 155–163 (2000)
199. Y.-F. Zheng, H. Hemami, Impact effects of biped contact with the environment. IEEE Trans. Syst. Man Cybern. **14**(3), 437–443 (1984)
200. V.V. Zhoga, Computation of walking robots movement energy expenditure, in *Proceedings of IEEE International Conference on Robotics and Automation* (Leuven, 1998), pp. 163–168
201. D. Zhou, K.H. Low, T. Zielinska, An efficient foot-force distribution algorithm for quadruped walking robots. Robotica **18**(4), 403–413 (2000)

Chapter 3
Kinematic Modeling and Analysis
of Six-Legged Robots

In this chapter, an integrated approach to carry out kinematic modeling and computer simulations of the motions and mechanisms of six-legged robots producing straight-forward, crab, and turning gaits are explained. New path planning approaches are suggested for the straight-forward, crab and turning motion analysis of the robot walking over any kind of terrain varying from flat to rough in three-dimensional Cartesian space with the desired gait pattern. Computer simulations have been carried out to test the effectiveness of the developed analytical model. The model is capable of investigating various kinematic parameters of the six-legged robot like displacement, velocity, acceleration, and trace of the position of aggregate center of mass.

3.1 Description of the Problem

The problem may be stated as follows: A six-legged robot has to plan its walking parameters during straight-forward, crab, and turning motions on various terrains in such a way that it can tackle those situations by consuming minimum amount of energy after ensuring optimal feet forces distributions and satisfying the conditions of static and dynamic stability.

3.1.1 Description of Proposed Six-Legged Walking Robot

In the present book, a 3D CAD model (Fig. 3.1) of an axisymmetric realistic six-legged robot is developed in a commercially available solid modeler named (CATIA V5) computer aided three-dimensional interactive application version 5 by executing the two main steps as described below [6]:

© Springer Nature Singapore Pte Ltd. 2020
A. Mahapatra et al., *Multi-body Dynamic Modeling of Multi-legged Robots*,
Cognitive Intelligence and Robotics, https://doi.org/10.1007/978-981-15-2953-5_3

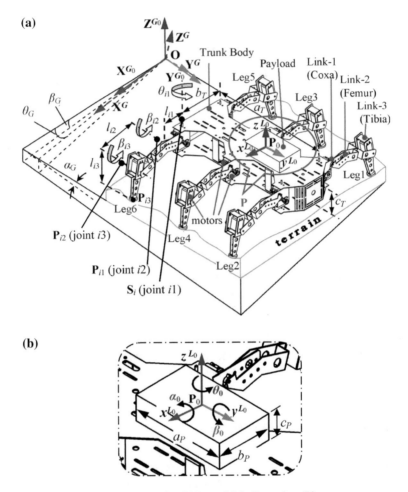

Fig. 3.1 A realistic six-legged robot **a** 3D CAD model, **b** Close view 'P'

Step 1: Modeling of the components of the robot in the part modeling workbench
Step 2: Assembling of the components in assembly workbench using suitable constraints.

The 3D CAD model as shown in Fig. 3.1 comprises of a trunk body (shape resembling a rectangle) and six identical legs. The three legs on either side of the central axis of the rectangular body are equally disseminated. Each leg is made of three links, namely coxa, femur, and tibia with their lengths being denoted by l_{ij}, where $i = 1$ to 6 represents the leg and $j = 1$ to 3 indicating the number of joint. The links are interconnected by two revolute joints (joints $i2$ and $i3$) and are connected to the trunk body by a third rotary joint (joint $i1$). Each joint of a leg is individually controlled by one actuator or DC servomotor. Hence, every single leg has three DOF which means with six legs of the robot there are eighteen DOF. These eighteen DOF corresponding to eighteen joint angles of the robot are required to work in a synchronized way to achieve the desired task. Hence, the spatial system can be considered to have a six DOF trunk body and eighteen DOF legs (=3DOF × 6legs), which collectively constitute a maximum of twenty-four mobility levels of the robotic system. The main components of the robot are made of aluminum (density: 2.74×10^{-6} kg/mm^3). The total mass of the robot without any payload is estimated to be equal to 2.456 kg.

3.1.2 Gait Terminologies and Their Relationships

Let us first review the following terms and their relationships with respect to wave gaits [5] of a six-legged robot:

- Gait: It is defined as the time and location for placing and lifting of each foot, coordinated with the motion of the body, in order to move the body from one place to another.
- Transfer or swing phase: Transfer phase of a leg is the period during which the foot is in air in a walking cycle.
- Support phase: It is the period during which the foot is placed on the ground in a walking cycle.
- Cycle time (T): It is the time for a complete cycle of leg's locomotion of a periodic gait.
- Duty factor (DF): It is the time fraction of a cycle time (T), during which a particular leg is in support phase.
- Stride (λ): It is the distance through which the center of gravity translates during one complete locomotion cycle.
- Body stroke (linear, s_0 or angular, s_0^c): It is the distance through which the trunk body moves during a complete support phase per cycle.
- Leg stroke (R_s): It is the distance through which the foot is translated relative to the trunk body during the support phase. Leg stroke must be within the leg workspace.
- Periodic gait: A gait is called periodic, if similar states of the same leg during successive strokes occur at the same interval for all legs, that interval being the cycle time. Otherwise, it is a non-periodic gait.
- Symmetric gait: A gait is called symmetric, if the motion of legs of any right-left pair is exactly half a cycle out of phase.
- Regular gait: It is a gait with the same duty factor for all the legs.
- Wave gait: It is a periodic, regular, and symmetric gait, in which the sequence of placing events of the legs on each side runs from the rear leg and proceeds toward the front leg. Therefore, the gait creates a wave of stepping from the rear to front.
- Crab axis: It is the axis, which goes through the body center and is aligned with the direction of body motion.
- Crab angle (θ_c): It is defined as the angle from the longitudinal axis of the robot body to the direction of motion.

- Lateral offset (l_i): It is the shortest distance between vertical projection of the hip on the ground and the corresponding track.
- Full swing (s_w^f): The distance traveled by the swing leg (end-to-end support) per cycle along the direction of motion.

3.2 Analytical Framework

The following assumptions are made in the present work:

- Robot executes motion planning in three-dimensional Cartesian space.
- A number of feet hold positions are already known before executing motion of the robot in various terrains.
- Predefined trunk body motion, swing leg motion, and slip trajectory are governed by cubic polynomials. This leads to variation in the trunk body height with respect to the various terrains throughout the motion.
- Slip occurs between the leg tip and the terrain during support phase and its magnitude is small. Further, the slip trajectory has been eased out to a straight path in the horizontal plane of the local frame of reference.

3.2.1 Reference System in Cartesian Coordinates

In this book, the robot is considered as a rigid multi-body system with multiple reference frames (both global and local frames) attached to it. Shown in Figs. 3.1 and 3.2, G_0 and G represent the static global and dynamic reference frames, respectively. The respective frames corresponding to *XYZ* coordinate systems coincide at **O**. The system has been generalized by means of orientation vector of *Bryant angles* [1], such that, (1) $\eta_G = [\alpha_G \ \beta_G \ \theta_G]^T$ denotes the orientation vector between G_0 and G as shown in Fig. 3.2. These angles take into account the terrain elevation along the three axes, i.e., α_G denotes sloping angle, β_G represents banking angle, and θ_G indicates initial angular position of the robot with respect to the frame G_0; (2) $\eta_0 = [\alpha_0 \ \beta_0 \ \theta_0]^T$ denotes the orientation vector between G and L_0 as shown in Fig. 3.1.

The reference frame L_0, fixed to the body, is affixed to an arbitrarily chosen location P_0 on the trunk body (refer to Fig. 3.3). The local reference frames are attached at each joint $i1$, $i2$, and $i3$ to signify successive joint states at any instant of time, that is, at time t (one state) and at time $t + \Delta t$ (successive state) after infinitesimally small rotation. More accurately, it can be said that frames L_i' and L_i'' are positioned at point S_i, frames L_{i1}', and L_{i1}'' at point P_{i1} and frames L_{i2}' and L_{i2}'' at point P_{i2}. Furthermore, frames L_{i3}' and L_{i3} (L_{i3} is considered parallel to frame G) are positioned at the tip point P_{i3}. This point follows an estimated trajectory in 3D Cartesian space (refer to [2]). To signify the crab angle (θ_c) and turning angle ($\theta_{i3}^{L_{i3}}$),

Fig. 3.2 References frames
G_0, G showing three
successive rotations

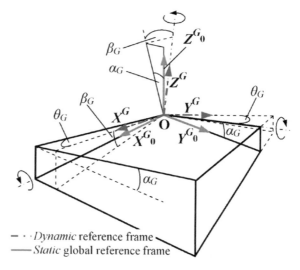

Fig. 3.3 Reference frames
and Vector diagram of leg 'i'
during support phase **a** plan
view, **b** elevation view

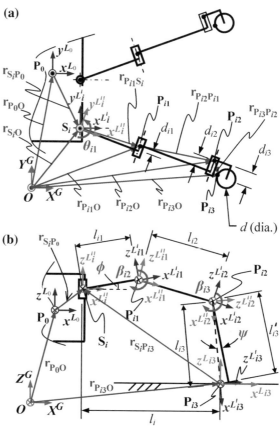

two more reference frames have been considered at point $\mathbf{P}_{i3}$, that is, $\mathbf{L}_{i3}''$, $\mathbf{L}_{i3}'''$. This has reference to Fig. 3.4b, c, respectively.

The displacement vectors represented in dynamic reference frame G from point $\mathbf{O}$ to $\mathbf{P}_0$, $\mathbf{S}_i$, $\mathbf{P}_{ij}$ (for $j = 1$ to 3) are given by $\mathbf{r}_{P_0 O}^G$, $\mathbf{r}_{S_i O}^G$, and $\mathbf{r}_{P_{ij} O}^G$, respectively. Similarly, the vectors of Cartesian coordinates of $\mathbf{P}_0$, $\mathbf{P}_{ij}$ with respect to G are represented, respectively, by,

$$\mathbf{p}_0^G = ((\mathbf{r}_{P_0 O}^G)^\mathrm{T}, \boldsymbol{\eta}_0^\mathrm{T}))^\mathrm{T} \in \mathbb{R}^6, \tag{3.1}$$

$$\mathbf{p}_{ij}^G = ((\mathbf{r}_{P_{ij} O}^G)^\mathrm{T}, \boldsymbol{\eta}_{ij}^\mathrm{T}))^\mathrm{T} \in \mathbb{R}^{108}, \tag{3.2}$$

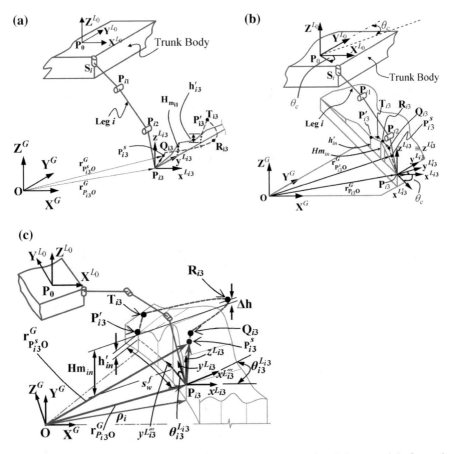

Fig. 3.4 Topography of terrain and swing leg trajectory planning for ith leg **a** straight-forward motion, **b** crab motion, **c** turning motion

where

$$\mathbf{r}_{P_0 O}^{G} = [\, x_{P_0 O}^{G} \;\; y_{P_0 O}^{G} \;\; z_{P_0 O}^{G} \,]^{\mathrm{T}}, \tag{3.3}$$

$$\boldsymbol{\eta}_0 = [\alpha_0 \;\; \beta_0 \;\; \theta_0]^{\mathrm{T}}, \tag{3.4}$$

$$\mathbf{r}_{P_{ij} O}^{G} = [\, x_{P_{ij} O}^{G} \;\; y_{P_{ij} O}^{G} \;\; z_{P_{ij} O}^{G} \,]^{\mathrm{T}}, \quad (\text{for } i = 1 \text{ to } 6, \; j = 1 \text{ to } 3) \tag{3.5}$$

$$\boldsymbol{\eta}_{ij} = [\alpha_{ij} \;\; \beta_{ij} \;\; \theta_{ij}]^{\mathrm{T}}, \quad (\text{for } i = 1 \text{ to } 6, \; j = 1 \text{ to } 3) \tag{3.6}$$

The displacement vector of the trunk body and the joints of each leg are first computed in reference to dynamic reference frame G. Thereafter, it is transformed to static global reference frame G_0 using the transform matrix, $\mathbf{A}^{G_0 G}$ (refer to Appendix A.1). Therefore, the vector of Cartesian coordinates of $\mathbf{P}_0$, $\mathbf{P}_{ij}$ with respect to G_0 are, respectively, given by,

$$\mathbf{p}_0^{G_0} = ((\mathbf{r}_{P_0 O}^{G_0})^{\mathrm{T}}, \boldsymbol{\eta}_0^{\mathrm{T}}))^{\mathrm{T}} \in \mathbb{R}^6, \tag{3.7}$$

$$\mathbf{p}_{ij}^{G_0} = ((\mathbf{r}_{P_{ij} O}^{G_0})^{\mathrm{T}}, \boldsymbol{\eta}_{ij}^{\mathrm{T}}))^{\mathrm{T}} \in \mathbb{R}^{108}, \tag{3.8}$$

Since every joint (i.e., joint $i1$, $i2$ and $i3$ as shown in Fig. 3.1) of leg 'i' has one DOF revolute joint,

$$\alpha_{i1} = \beta_{i1} = \alpha_{i2} = \theta_{i2} = \alpha_{i3} = \theta_{i3} = 0, \tag{3.9}$$

that is, the *Bryant angles* (Eq. 3.6) can be further reduced to,

$$\boldsymbol{\eta}_{i1} = [0 \; 0 \; \theta_{i1}]^{\mathrm{T}}, \tag{3.10}$$

$$\boldsymbol{\eta}_{i2} = [0 \; \beta_{i2} \; 0]^{\mathrm{T}}, \tag{3.11}$$

$$\boldsymbol{\eta}_{i3} = [0 \; \beta_{i3} \; 0]^{\mathrm{T}}, \tag{3.12}$$

respectively, by following *Z-Y-Y* convention.

The kinematic diagram of ith leg of the realistic six-legged robot is shown in Fig. 3.3. Here, ϕ denotes the twisted angle of the coxa (in the present book, $\phi = 0$), l_i is the distance between the projection of the points $\mathbf{S}_i$ and $\mathbf{P}_{i3}$ on the plane XY of frame G commonly termed as the lateral offset, d is the diameter of the footpad, l'_{i3} is the distance between joint $i3$ and footpad center. The offset distance between the links of the ith leg is denoted by d_{i1}, $d_{i2,}$ and d_{i3}, respectively.

From the geometrical configuration shown in Fig. 3.3, the subsequent relations are achieved:

Table 3.1 Physical parameters of the robot

Components	Effective dimension (m)	Mass (kg)
Trunk Body	$a_T = 0.420; b_T = 0.160; c_T = 0.090$	0.650
Payload	$a_P = 0.150; b_P = 0.090; c_P = 0.040$	4.244
Link $i1$	$l_{i1} = 0.085$	0.150
Link $i2$	$l_{i2} = 0.120$	0.041
Link $i3$	$l_{i3} = 0.100$	0.110

NB: In the present book, twisted angle of the coxa, $\phi = 0$ deg; $l_i = 218.5$ mm; $d_{i1} = 8$ mm; $d_{i2} = 17.75$ mm; $d_{i3} = 19.75$ mm; $d = 25$ mm; a_T, b_T, and c_T are dimensions of the trunk body as shown in Fig. 3.1a. Similarly, a_P, b_P, and c_P are dimensions of the payload as shown in Fig. 3.1b

Bounding angle,

$$\psi = \tan^{-1}\left(d / 2l'_{i3}\right), \tag{3.13}$$

Kinematic link length,

$$l_{i3} = l'_{i3} / \cos\psi, \tag{3.14}$$

The ranges of the joint variables: $\theta_{i1}, \beta_{i2}, \beta_{i3}$ for turning motions are limited by the geometrical structure of the robot. Table 3.1 shows the physical parameters of the realistic hexapedal robot.

3.2.2 Kinematic Constraint Equations

The kinematic constraints of the multi-legged robotic system relate the orientation and position of the trunk body, joint ij ($i = 1$ to 6, $j = 1$ to 3) and foot tip (in both support and swing phase) with respect to frame G, which are ultimately transformed to frame G_0 using the transform $\mathbf{A}^{G_0 G}$. As discussed in Sect. 3.1.1, the trunk body consists of six DOF. Every leg consists of three rotary joints located at $\mathbf{S}_i$, $\mathbf{P}_{i1}$ and $\mathbf{P}_{i2}$, while each of the feet tips is assumed to be spherical joint in stance phase located at $\mathbf{P}_{i3}$ (refer to Fig. 3.3). In the following section, the *vector loop* and *orientation loop* equations [1] depicting the orientation and position of $\mathbf{P}_0$, $\mathbf{S}_i$, $\mathbf{P}_{ij}$ of leg i are considered with respect to frame G. Refer to Table 3.2.

Now, the overall set of explicitly constrained position equations that governs the system at any instant are represented by the functions,

(a) With respect to frame G,

$$\mathbf{g}(\mathbf{p}^G) = 0 \quad \in \mathbb{R}^{114} \tag{3.15}$$

Table 3.2 Constraint loop equations (Refer to Appendix A.2)

Loop	Vector loop equation	Joint	Orientation loop equation
$OP_0 S_i O$	$\mathbf{r}^G_{P_0 O} + \mathbf{r}^G_{S_i P_0} - \mathbf{r}^G_{S_i O} = \mathbf{0}_3$ or $\mathbf{r}^G_{P_0 O} + \mathbf{A}^{GL_0}\mathbf{r}^{L_0}_{S_i P_0} - \mathbf{r}^G_{S_i O} = \mathbf{0}_3$	$\mathbf{S}_i$ [Joint $\mathbf{J}_{i1}$]	$\underbrace{\mathbf{A}^{GL''_i}}_{\substack{\text{free}\\\text{rotation}}} \cdot \underbrace{\mathbf{A}^{L''_i L'_i}}_{\substack{\text{free}\\\text{rel.vel.}}} \cdot \underbrace{\mathbf{A}^{L'_i L_0}}_{\substack{\text{fixed}\\\text{rotation}}} \cdot \underbrace{\mathbf{A}^{L_0 G}}_{\substack{\text{free}\\\text{rotation}}} = \mathbf{I}_3$ or $\mathbf{A}^{GL''_i} = \mathbf{A}^{GL_0}.\mathbf{A}^{L_0 L'_i}.\mathbf{A}^{L'_i L''_i}$
$OS_i P_{i1} O$	$\mathbf{r}^G_{S_i O} + \mathbf{r}^G_{P_{i1} S_i} - \mathbf{r}^G_{P_{i1} O} = \mathbf{0}_3$ or $\mathbf{r}^G_{S_i O} + \mathbf{A}^{GL''_i}\mathbf{r}^{L''_i}_{P_{i1} S_i} - \mathbf{r}^G_{P_{i1} O} = \mathbf{0}_3$	$\mathbf{P}_{i1}$ [Joint $\mathbf{J}_{i2}$]	$\underbrace{\mathbf{A}^{GL''_{i1}}}_{\substack{\text{free}\\\text{rotation}}} \cdot \underbrace{\mathbf{A}^{L''_{i1} L'_{i1}}}_{\substack{\text{free}\\\text{rel.vel.}}} \cdot \underbrace{\mathbf{A}^{L'_{i1} L''_i}}_{\substack{\text{fixed}\\\text{rotation}}} \cdot \underbrace{\mathbf{A}^{L''_i G}}_{\substack{\text{free}\\\text{rotation}}} = \mathbf{I}_3$ or $\mathbf{A}^{GL''_{i1}} = \mathbf{A}^{GL''_i}.\mathbf{A}^{L''_i L'_{i1}}.\mathbf{A}^{L'_{i1} L''_{i1}}$
$OP_{i1} P_{i2} O$	$\mathbf{r}^G_{P_{i1} O} + \mathbf{r}^G_{P_{i2} P_{i1}} - \mathbf{r}^G_{P_{i2} O} = \mathbf{0}_3$ or $\mathbf{r}^G_{P_{i1} O} + \mathbf{A}^{GL''_{i1}}\mathbf{r}^{L''_{i1}}_{P_{i2} P_{i1}} - \mathbf{r}^G_{P_{i2} O} = \mathbf{0}_3$	$\mathbf{P}_{i2}$ [Joint $\mathbf{J}_{i3}$]	$\underbrace{\mathbf{A}^{GL''_{i2}}}_{\substack{\text{free}\\\text{rotation}}} \cdot \underbrace{\mathbf{A}^{L''_{i2} L'_{i2}}}_{\substack{\text{free}\\\text{rel.vel.}}} \cdot \underbrace{\mathbf{A}^{L'_{i2} L''_{i1}}}_{\substack{\text{fixed}\\\text{rotation}}} \cdot \underbrace{\mathbf{A}^{L''_{i1} G}}_{\substack{\text{free}\\\text{rotation}}} = \mathbf{I}_3$ or $\mathbf{A}^{GL''_{i2}} = \mathbf{A}^{GL''_{i1}}.\mathbf{A}^{L''_{i1} L'_{i2}}.\mathbf{A}^{L'_{i2} L''_{i2}}$
$OP_{i2} P_{i3} O$	$\mathbf{r}^G_{P_{i2} O} + \mathbf{r}^G_{P_{i3} P_{i2}} - \mathbf{r}^G_{P_{i3} O} = \mathbf{0}_3$ or $\mathbf{r}^G_{P_{i2} O} + \mathbf{A}^{GL''_{i2}}\mathbf{r}^{L''_{i2}}_{P_{i3} P_{i2}} - \mathbf{r}^G_{P_{i3} O} = \mathbf{0}_3$ $\mathbf{r}^G_{P_{i3} O} - \mathbf{r}^G_i = \mathbf{0}_3$		

(b) With respect to frame G_0

$$\mathbf{g}(\mathbf{p}^{G_0}) = 0 \quad \in \mathbb{R}^{114} \qquad (3.16)$$

The function $\mathbf{g}(\mathbf{p}^G)$ or $\mathbf{g}(\mathbf{p}^{G_0})$ has both kinematic joint constraints and driving constraints that results into mobility (m_b) of the system [4], as given by

$$m_b = 6n_b - n_c, \qquad (3.17)$$

where n_b is the total number of rigid bodies of the system and n_c is the number of kinematic joint constraints. The robotic system in this book has only revolute joints. Therefore, a simple formula has been derived such that,

$$n_c = 5(n_b - 1) + 3n_g \tag{3.18}$$

Here, n_g is the number of grounded legs.

In the present system, there are 108 kinematic joint constraint equations, if all the legs are in support, 99 kinematic joint constraint equations, if three legs are in support, 102 kinematic joint constraint equations, if four legs are in support, and so on. The vector of Cartesian coordinates are represented by,

$$\mathbf{p}^G = ((\mathbf{p}_0^G)^{\mathrm{T}}, (\mathbf{p}_1^G)^{\mathrm{T}}, (\mathbf{p}_2^G)^{\mathrm{T}}, (\mathbf{p}_3^G)^{\mathrm{T}}, (\mathbf{p}_4^G)^{\mathrm{T}}, (\mathbf{p}_5^G)^{\mathrm{T}}, (\mathbf{p}_6^G)^{\mathrm{T}})^{\mathrm{T}} \quad \in \mathbb{R}^{114}, \tag{3.19}$$

$$\mathbf{p}^{G_0} = ((\mathbf{p}_0^{G_0})^{\mathrm{T}}, (\mathbf{p}_1^{G_0})^{\mathrm{T}}, (\mathbf{p}_2^{G_0})^{\mathrm{T}}, (\mathbf{p}_3^{G_0})^{\mathrm{T}}, (\mathbf{p}_4^{G_0})^{\mathrm{T}}, (\mathbf{p}_5^{G_0})^{\mathrm{T}}, (\mathbf{p}_6^{G_0})^{\mathrm{T}})^{\mathrm{T}} \quad \in \mathbb{R}^{114}, \tag{3.20}$$

$$\mathbf{p}_i^G = ((\mathbf{p}_{i1}^G)^{\mathrm{T}}, (\mathbf{p}_{i2}^G)^{\mathrm{T}}, (\mathbf{p}_{i3}^G)^{\mathrm{T}})^{\mathrm{T}} \in \mathbb{R}^{18}, \tag{3.21}$$

$$\mathbf{p}_i^{G_0} = ((\mathbf{p}_{i1}^{G_0})^{\mathrm{T}}, (\mathbf{p}_{i2}^{G_0})^{\mathrm{T}}, (\mathbf{p}_{i3}^{G_0})^{\mathrm{T}})^{\mathrm{T}} \in \mathbb{R}^{18}. \tag{3.22}$$

The vector of Cartesian coordinates $\mathbf{p}_{ij}^G$ and $\mathbf{p}_{ij}^{G_0}$ for ($i = 1$ to 6, $j = 1$ to 3) can obtained from Eqs. (3.2) and (3.8), respectively. Therefore, the holonomic constraint equations governing the state of the trunk body and ith leg of the robotic system are depicted by a set of equations, as shown in Eq. (3.23).

$$\mathbf{g}_i(p_i^G) = \begin{pmatrix} \mathbf{g}_1(\mathbf{p}_0^G) \\ \mathbf{g}_2(\mathbf{p}_0^G) \\ \mathbf{g}_{3i}(\mathbf{p}_i^G) \\ \mathbf{g}_{4i}(\mathbf{p}_i^G) \\ \mathbf{g}_{5i}(\mathbf{p}_i^G) \\ \mathbf{g}_{6i}(\mathbf{p}_i^G) \\ \mathbf{g}_{7i}(\mathbf{p}_i^G) \\ \mathbf{g}_{8i}(\mathbf{p}_i^G) \\ \mathbf{g}_{9i}(\mathbf{p}_i^G) \end{pmatrix} \equiv \begin{pmatrix} \mathbf{r}_{P_0 O}^G - \int \mathbf{f}(t) \\ \mathbf{\eta}_0 - \int \mathbf{g}(t) \\ \mathbf{r}_{P_0 O}^G + \mathbf{A}^{GL_0} \mathbf{r}_{S_i P_0}^{L_0} + \mathbf{A}^{GL_i''} \mathbf{r}_{P_{i1} S_i}^{L_i''} - \mathbf{r}_{P_{i1} O}^G \\ \mathbf{P}_r^{\mathrm{T}}(x, y).\mathbf{A}^{L_i'' G}.\mathbf{A}^{GL_i'}.\mathbf{P}_r(z) \\ \mathbf{r}_{P_{i1} O}^G + \mathbf{A}^{GL_{i1}''} \mathbf{r}_{P_{i2} P_{i1}}^{L_{i1}''} - \mathbf{r}_{P_{i2} O}^G \\ \mathbf{P}_r^{\mathrm{T}}(x, z).\mathbf{A}^{L_{i1}'' G}.\mathbf{A}^{GL_{i1}'}.\mathbf{P}_r(y) \\ \mathbf{r}_{P_{i2} O}^G + \mathbf{A}^{GL_{i2}''} \mathbf{r}_{P_{i3} P_{i2}}^{L_{i2}''} - \mathbf{r}_{P_{i3} O}^G \\ \mathbf{P}_r^{\mathrm{T}}(x, z).\mathbf{A}^{L_{i2}'' G}.\mathbf{A}^{GL_{i2}'}.\mathbf{P}_r(y) \\ \mathbf{r}_{P_{i3} O}^G - \mathbf{r}_{P_{i3}^s O}^G \end{pmatrix} = \begin{pmatrix} \mathbf{0}_3 \\ \mathbf{0}_3 \\ \mathbf{0}_3 \\ \mathbf{0}_2 \\ \mathbf{0}_3 \\ \mathbf{0}_2 \\ \mathbf{0}_3 \\ \mathbf{0}_2 \\ \mathbf{0}_3 \end{pmatrix} \tag{3.23}$$

Further, the present problem is about solving the joint displacements, velocities, and accelerations for a selected trunk body motion and swing leg trajectory for a specified gait on various grounds. $\mathbf{g}_1(\mathbf{p}_0^G)$ and $\mathbf{g}_2(\mathbf{p}_0^G)$ that denotes the trunk body motion constraints and $\mathbf{g}_{9i}(\mathbf{p}_i^G)$ denoting swing leg motion constraints are considered as the driving constraints in the present problem $\mathbf{P}_r^{\mathrm{T}}(x, y)$, $\mathbf{P}_r^{\mathrm{T}}(x, z)$, $\mathbf{P}_r(z)$, and $\mathbf{P}_r(y)$ are mainly the matrix projectors (refer to Appendix A.1). Further, $\mathbf{f}(t)$ and $\mathbf{g}(t)$, the translational and angular constraint derivatives governing the trunk body motion at given instant of time, are detailed in Sect. 3.2.5.1. Also, due to trajectory planning (refer to Sect. 3.2.5.2) and gait planning (refer to Sect. 3.2.6) of leg i, $\mathbf{r}_{P_{i3}^s O}^G$ is the

Table 3.3 Coordinates of the local components of the robotic system (in mm)

Location	Leg 2, Leg 4, Leg 6	Leg 1, Leg 3, Leg 5
S_i	$\mathbf{r}_{S_2 P_0}^{L_0} = (80.0, 191.3, 23.5)^{\mathrm{T}}$	$\mathbf{r}_{S_1 P_0}^{L_0} = (-80.0, 191.3, 23.5)^{\mathrm{T}}$
	$\mathbf{r}_{S_4 P_0}^{L_0} = (80.0, -18.7, 23.5)^{\mathrm{T}}$	$\mathbf{r}_{S_3 P_0}^{L_0} = (-80.0, -18.7, 23.5)^{\mathrm{T}}$
	$\mathbf{r}_{S_6 P_0}^{L_0} = (80.0, -228.7, 23.5)^{\mathrm{T}}$	$\mathbf{r}_{S_5 P_0}^{L_0} = (-80.0, -228.7, 23.5)^{\mathrm{T}}$
$\mathbf{P}_{i1}$	$\mathbf{r}_{P_{i1} S_i}^{L_i''} = (l_{i1}.c_\phi, d_{i1}, l_{i1}.s_\phi)^{\mathrm{T}}$	$\mathbf{r}_{P_{i1} S_i}^{L_i''} = (-l_{i1}.c_\phi, d_{i1}, l_{i1}.s_\phi)^{\mathrm{T}}$
$\mathbf{P}_{i2}$	$\mathbf{r}_{P_{i2} P_{i1}}^{L_{i1}''} = (l_{i2}, d_{i2}, 0)^{\mathrm{T}}$	$\mathbf{r}_{P_{i2} P_{i1}}^{L_{i1}''} = (-l_{i2}, d_{i2}, 0)^{\mathrm{T}}$
$\mathbf{P}_{i3}$	$\mathbf{r}_{P_{i3} P_{i2}}^{L_{i2}''} = (l_{i3}, -d_{i3}, 0)^{\mathrm{T}}$	$\mathbf{r}_{P_{i3} P_{i2}}^{L_{i2}''} = (-l_{i3}, -d_{i3}, 0)^{\mathrm{T}}$

coordinates of the tip point of link l_{i3} at any instant of time (both the stance and swing phases). The local components of $\mathbf{S}_i$, $\mathbf{P}_{i1}$, $\mathbf{P}_{i2}$ and $\mathbf{P}_{i3}$ are fixed for a robotic structure and are given by $\mathbf{r}_{S_i P_0}^{L_0}$, $\mathbf{r}_{P_{i1} S_i}^{L_i''}$, $\mathbf{r}_{P_{i2} P_{i1}}^{L_{i1}''}$, $\mathbf{r}_{P_{i3} P_{i2}}^{L_{i2}''}$, respectively, (refer to Table 3.3 for the specified coordinates of robot). Also, $\mathbf{A}^{GL_0}$, $\mathbf{A}^{GL_i'}$, $\mathbf{A}^{GL_i''}$, $\mathbf{A}^{GL_{i1}''}$ $\mathbf{A}^{GL_{i1}''}$, $\mathbf{A}^{GL_{i2}''}$, $\mathbf{A}^{GL_{i2}''}$ are the transformation matrices related to different local frames with respect to frame G. $\mathbf{A}^{L_i''G}$, $\mathbf{A}^{L_{i1}''G}$, $\mathbf{A}^{L_{i2}''G}$ are orthogonal matrices of $\mathbf{A}^{GL_i''}$, $\mathbf{A}^{GL_{i1}''}$, $\mathbf{A}^{GL_{i2}''}$, respectively. The transformation matrices are mentioned in Appendix A.3.

3.2.3   Inverse Kinematic Model of the Six-Legged Robotic System

It is necessary to calculate the explicit inverse kinematic solutions of the system for controlling a six-legged robot in 3D Cartesian space, which basically refers to determination of joint angles of the leg i (for $i = 1$ to 6) for the specified trunk body and leg-tip displacement of leg i at given instant of time.

First of all, the joint variables θ_{i1}, β_{i2}, β_{i3} (in both stance and swing phases) of leg i of the six-legged robot are determined from the inverse kinematic model and subsequently, the velocities and accelerations (both linear and angular) are calculated. In the support phase (shown in Fig. 3.4), the position vector of $\mathbf{S}_i$ with respect to frame G and frame L_{i3}' can be determined by using *vector loop* equations, respectively, as depicted by

$$\mathbf{r}_{S_i P_{i3}}^{G} = \mathbf{r}_{P_0 O}^{G} + \mathbf{r}_{S_i P_0}^{G} - \mathbf{r}_{P_{i3} O}^{G} \tag{3.24}$$

and

$$\mathbf{r}_{S_i P_{i3}}^{L_{i3}'} = -\mathbf{r}_{P_{i3} P_{i2}}^{L_{i3}'} - \mathbf{r}_{P_{i2} P_{i1}}^{L_{i3}'} - \mathbf{r}_{P_{i1} S_i}^{L_{i3}'} \tag{3.25}$$

It is to be noted that: $\mathbf{r}^G_{P_0 O}$ and $\mathbf{r}^G_{P_{i3} O}$, the position vectors, are time dependent and, respectively, govern the motions of trunk body and swing leg's trajectory with respect to the frame G. Further, joint angles at any instant of time for locomotion analysis of the robot over various terrains are computed from the developed inverse kinematic model with suitable boundary conditions (predefined time derivative functions of the trajectories of both the trunk body and swing legs).

Equation (3.24) can also be written as follows:

$$\mathbf{r}^G_{S_i P_{i3}} = \mathbf{A}^{G L'_{i3}} . \mathbf{r}^{L'_{i3}}_{S_i P_{i3}} = (a_i, \ b_i, \ c_i)^{\mathrm{T}} (\text{say}), \tag{3.26}$$

where a_i, b_i, c_i denote the coordinates of $\mathbf{r}^G_{S_i P_{i3}}$ and

$$\mathbf{A}^{G L'_{i3}} = \mathbf{A}^{G L''_{i2}} = \mathbf{A}^{G L_0} . \mathbf{A}^{L_0 L''_{i2}} \tag{3.27}$$

(frames L'_{i3} and L''_{i2} are parallel)

It is to be considered that $\mathbf{A}^{L_0 G}$ and $\mathbf{A}^{G L_0}$ are orthogonal transformation matrices calculated using transformations given in Appendix A.3.

Now, rearranging Eqs. (3.26) and (3.27) to obtain the governing equation of the system as follows:

$$\mathbf{A}^{L_0 G} . \mathbf{r}^G_{S_i P_{i3}} = \mathbf{A}^{L_0 L''_{i2}} . \mathbf{r}^{L'_{i3}}_{S_i P_{i3}} \tag{3.28}$$

The joint angles are obtained through method of elimination using suitable transformation matrices (refer Appendix A.3) and coordinates of the local components. They are summarized as follows:

$$\theta_{i1} = \gamma - 2n\pi - 2 \tan^{-1}((k_{i1} \pm k_{i4}) / (d_i + k_{i2})), \quad n \in I \tag{3.29}$$

$$\beta_{i2} = \phi - 2n\pi - 2 \tan^{-1}((k_{i6} \pm k_{i7}) / k_{i8}), \quad n \in I \tag{3.30}$$

$$\beta_{i3} = 2n\pi \pm 2 \tan^{-1} \sqrt{(1 - k_{i5}) / (1 + k_{i5})}, \quad n \in I \tag{3.31}$$

where

$$d_i = d_{i1} + d_{i2} - d_{i3} \tag{3.32}$$

$$k_{i1} = -(a_i (\cos \beta_0 \cos \theta_0) + b_i (\cos \alpha_0 \sin \theta_0 + \sin \alpha_0 \sin \beta_0 \cos \theta_0) + c_i (\sin \alpha_0 \sin \theta_0 - \cos \alpha_0 \sin \beta_0 \cos \theta_0)) \tag{3.33}$$

$$k_{i2} = -(-a_i(\cos\beta_0\sin\theta_0) + b_i(\cos\alpha_0\cos\theta_0 - \sin\alpha_0\sin\beta_0\sin\theta_0)$$
$$+ c_i(\sin\alpha_0\cos\theta_0 + \cos\alpha_0\sin\beta_0\sin\theta_0)) \tag{3.34}$$

$$k_{i3} = -(a_i(\sin\beta_0) - b_i(\sin\alpha_0\cos\beta_0) + c_i(\cos\alpha_0\cos\beta_0)) \tag{3.35}$$

$$k_{i4} = \sqrt{k_{i1}^2 + k_{i2}^2 - d_i^2} \tag{3.36}$$

$$k_{i5} = \left((k_{i3} - l_{i1}.\sin\phi)^2 + (k_{i4} - l_{i1}.\cos\phi)^2 - l_{i2}^2 - l_{i3}^2\right)\big/2l_{i2}.l_{i3} \tag{3.37}$$

$$k_{i6} = (k_{i3} - l_{i1}.s\phi) \tag{3.38}$$

$$k_{i7} = \sqrt{(k_{i3} - l_{i1}.s\phi)^2 + (k_{i4} - l_{i1}.c\phi)^2 - (l_{i2} + l_{i3}.k_{i5})^2} \tag{3.39}$$

$$k_{i8} = l_{i2} + l_{i3}.k_{i5} + k_{i4} - l_{i1}.c\phi \tag{3.40}$$

To compute the swing phase angles, it is necessary to substitute $k'_{i1} = -k_{i1}$, $k'_{i2} = -k_{i2}$ and $k'_{i3} = -k_{i3}$.

3.2.4 Terrain Model

In the current book, the topology of the terrain along the path of robot's locomotion varies from smooth to irregular that includes flat plain, slope, staircase, banking, undulation, etc. It is to be noted that in the present work, the data points of the topology are already predefined. Further, the kinematics of the robotic system can be intimately connected to the represented terrain in two methods.

Method I: In this method, the robot performs foot placement at ideal position, which results from the path (both trunk body and legs) and gait planning. The two points that are considered with respect to frame $\mathbf{I}_{i3}$, namely $\mathbf{P}_{i3}$ (starting point of swing) and $\mathbf{P}'_{i3}$ (subsequent end points of swing), defines the topology of the terrain. The terrain height denoted by h'_{in} (i is the leg number and n is the duty cycle number) is measured along Z-axis with respect to frame $\mathbf{L}_{i3}$ (as shown in Fig. 3.4) such that the foot tip of swing leg i lands on the terrain in the nth cycle. The maximum terrain height denoted by Hm_{in} (i is the leg number and n is the duty cycle number) and is taken along Z-axis with respect to frame $\mathbf{L}_{i3}$ and is basically the height of the terrain along the path of swing in the nth cycle. Finally, the data points computed with respect to frame $\mathbf{L}_{i3}$ are transformed to reference frame $\mathbf{G}$ using suitable transformation matrices. The values of h'_{in} and Hm_{in} are provided as inputs for computer simulations.

Method II: This method mainly deals with the issue of map-based foot placement and subsequently, path generation. Based on the availability of the topographical information of the terrain (given as input), the robot can carry out its path planning from initial to final position. The three-dimensional CAD model of the topography is developed and the data points are extracted from it with respect to the reference frame G. During simulation, maximum height (Hm_{in}) of the terrain which the swing leg has to overcome along its path in the workspace is calculated using the developed algorithm. Moreover, the terrain height h'_{in} with respect to local frame L_{i3}, that is, $z^G_{P_{i3}O}$ with respect to frame G in support phase is determined based on the information of the coordinates $x^G_{P_{i3}O}$ and $y^G_{P_{i3}O}$. This means the foot tip of leg i searches for the possible foot placement location during each time step and subsequently calculates the coordinate of the location that assists the robot to continue an efficient gait.

In both the methods, it is assumed that the maximum height of the swing trajectory is always greater than the maximum height of topography through which the leg traverses and the difference is given by Δh.

3.2.5 Locomotion Planning on Various Terrains

In the current book, attempt has been made to make the motion characteristics of the legged robot (both trunk body and swing legs) more realistic compare to the previous studies by various researchers worldwide. For that motion of the trunk body (for location $\mathbf{P_0}$) and leg tip of swing leg (for location $\mathbf{P_{i3}}$) are modeled based on Automated Dynamic Analysis of Mechanical Systems (ADAMS) step function [3]. The function approximates the Heaviside step function with a cubic polynomial. It is given by the following expression:

$$h = h_a + a.\Delta^2(3 - 2\Delta), \quad \text{for} \quad t_a \text{ to } t_b \tag{3.41}$$

where

$$a = h_b - h_a, \tag{3.42}$$

($h_a =$ initial step value at time t_a, $h_b =$ final step value at time t_b)

$$\Delta = (t - t_a)/(t_b - t_a) \tag{3.43}$$

Equation (3.41) has a smooth function value during the transition over a specified interval of an independent variable. The function has a continuous first derivative with smooth changes at the transition point which makes it the basis for consideration in this book.

3.2.5.1 Motion Planning for Robot's Body

The motion planning of the trunk body of the robot should be such that there is an uninterrupted and continuous motion for the given initial position, orientation (roll, pitch, and yaw), and maximum rate of change of angular displacement of the trunk body. In this book, the angular and translational constraint derivatives are given by $\mathbf{g}(t) = [g_x(t) \ g_y(t) \ g_z(t)]^T$ and $\mathbf{f}(t) = [f_x(t) \ f_y(t) \ f_z(t)]^T$, respectively. The six driving constraint equations representing the six degrees of freedom of the trunk body govern the state of the inverse kinematics problem. Assumption of such driving constraints makes the locomotion of the trunk body more realistic, such that the body can be raised, lowered, or tilted with the legs' movements in a synchronized way. This is much necessary for a robot to maneuver in an irregular terrain with stable gaits strategy. It is to be noted that these constraints are assumed to follow a cubic polynomial [refer to (Eq. 3.41)] as discussed earlier.

(a) *Calculation of angular displacement rate inputs*

   Mathematically,

$$\dot{\boldsymbol{\eta}}_0 = \mathbf{g}(t) \in \mathbb{R}^{108} \tag{3.44}$$

Along X-direction (Pitching):
The rate of change of angular displacement [i.e., $g_x(t)$] of the robot for three cycles along X-direction of frame G is assumed to follow a pattern as shown in Fig. 3.5. The mathematical expression is given as follows:

   Cycle-1

$$\begin{aligned} g_x(t) &= \dot{\alpha}_0|_{t=t_0} + a_{\dot{\alpha}_0}|_1 . \Delta_{as}|_1^2 . (3 - 2\Delta_{as}|_1) \quad \text{for } t_0 \text{ to } t_1|_1 \\ &= \dot{\alpha}_0|_{t=t_1|_1} \qquad\qquad\qquad\qquad\qquad \text{for } t = t_1|_1 \\ &= \dot{\alpha}_0|_{t=t_1|_1} - a_{\dot{\alpha}_0}|_1 . \Delta_{ds}|_1^2 . (3 - 2\Delta_{ds}|_1) \quad \text{for } t_1|_1 \text{ to } t_3^s|_1 \end{aligned} \tag{3.45}$$

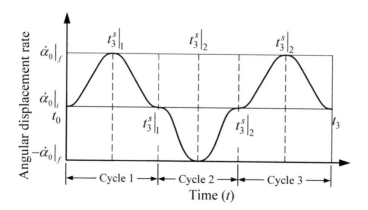

Fig. 3.5 Angular displacement rate versus time

where

$$
\begin{aligned}
a_{\dot{\alpha}_0}\big|_1 &= \dot{\alpha}_0\big|_{t=t_1|_1} - \dot{\alpha}_0\big|_{t=t_0}; \\
\Delta_{as}\big|_1 &= (t - t_0)\big/(t_1|_1 - t_0); \\
\Delta_{ds}\big|_1 &= (t - t_1|_1\big/(t_3^s\big|_1 - t_1|_1).
\end{aligned}
\tag{3.46}
$$

Here, $\dot{\alpha}_0\big|_{t=t_0}$ and $\dot{\alpha}_0\big|_{t=t_1|_1}$ are angular displacement rates of the trunk body along X-direction of frame G at time $t = t_0$ and $t = t_1|_1$, respectively; $t_3^s\big|_1$ is the end time of Cycle-1. Similarly,

Cycle-2

$$
\begin{aligned}
g_x(t) &= \dot{\alpha}_0\big|_{t=t_3^s|_1} - a_{\dot{\alpha}_0}|_2.\Delta_{as}|_2^2.(3 - 2\Delta_{as}|_2) \quad \text{for } t_3^s\big|_1 \text{ to } t_1|_2 \\
&= -\dot{\alpha}_0\big|_{t=t_1|_2} \qquad\qquad\qquad\qquad\quad \text{for } t = t_1|_2 \\
&= -\dot{\alpha}_0\big|_{t=t_1|_2} + a_{\dot{\alpha}_0}|_2.\Delta_{ds}|_2^2.(3 - 2\Delta_{ds}|_2) \quad \text{for } t_1|_2 \text{ to } t_3^s\big|_2
\end{aligned}
\tag{3.47}
$$

where

$$
\begin{aligned}
a_{\dot{\alpha}_0}\big|_2 &= \dot{\alpha}_0\big|_{t=t_1|_2} - \dot{\alpha}_0\big|_{t=t_3^s|_1}; \\
\Delta_{as}\big|_2 &= (t - t_3^s\big|_1)\big/(t_1|_2 - t_3^s\big|_1); \\
\Delta_{ds}\big|_2 &= (t - t_1|_2)\big/(t_3^s\big|_2 - t_1|_2).
\end{aligned}
\tag{3.48}
$$

Here, $\dot{\alpha}_0\big|_{t=t_3^s|_1}$ and $\dot{\alpha}_0\big|_{t=t_1|_2}$ are angular displacement rates of the trunk body along X-direction at time $t = t_3^s\big|_1$ and $t = t_1|_2$, respectively; $t_3^s\big|_2$ is the end time of Cycle-2. Similarly,

Cycle-3

$$
\begin{aligned}
g_x(t) &= \dot{\alpha}_0\big|_{t=t_3^s|_2} + a_{\dot{\alpha}_0}\big|_3.\Delta_{as}|_3^2.(3 - 2\Delta_{as}|_3) \quad \text{for } t_3^s\big|_2 \text{ to } t_1|_3 \\
&= \dot{\alpha}_0\big|_{t=t_1|_3} \qquad\qquad\qquad\qquad\quad \text{for } t = t_1|_3 \\
&= \dot{\alpha}_0\big|_{t=t_1|_3} - a_{\dot{\alpha}_0}\big|_3.\Delta_{ds}|_3^2.(3 - 2\Delta_{ds}|_3) \quad \text{for } t_1|_3 \text{ to } t_3
\end{aligned}
\tag{3.49}
$$

where

$$
\begin{aligned}
a_{\dot{\alpha}_0}\big|_3 &= \dot{\alpha}_0\big|_{t=t_1|_3} - \dot{\alpha}_0\big|_{t=t_3^s|_2}; \\
\Delta_{as}\big|_3 &= (t - t_3^s\big|_2)\big/(t_1|_3 - t_3^s\big|_2); \\
\Delta_{ds}\big|_3 &= (t - t_1|_3)\big/(t_3 - t_1|_3).
\end{aligned}
\tag{3.50}
$$

Here, $\dot{\alpha}_0\big|_{t=t_3^s|_2}$ and $\dot{\alpha}_0\big|_{t=t_1|_3}$ are angular displacement rates of the trunk body along X-direction at time $t = t_3^s\big|_2$ and $t = t_1|_3$, respectively; t_3 is the total cycle time for Cycle-3.

Along Y-direction (Rolling):
The same procedure as above is followed in the case of $g_y(t)$ having angular orientation angle β_0.

Along Z (Yawing) only during TURNING Motion:
The governing function for yaw during turning, i.e., the rate of change of angular displacement [i.e., $gz(t)$] of the trunk body at any instant of time along Z-axis of frame G is given by,

$$
\begin{aligned}
g_z(t) &= \dot{\theta}_0\big|_{t=t_0} + a_{\dot{\theta}_0}\Delta_a^2(3 - 2\Delta_a) && \text{for } t_0 \text{ to } t_1 \\
&= \dot{\theta}_0\big|_{t=t_1} && \text{for } t_1 \text{ to } t_2 \\
&= \dot{\theta}_0\big|_{t=t_2} - a_{\dot{\theta}_0}\Delta_d^2(3 - 2\Delta_d) && \text{for } t_2 \text{ to } t_3
\end{aligned}
\tag{3.51}
$$

where

$$
\begin{aligned}
a_{\dot{\theta}_0} &= \dot{\theta}_0\big|_{t=t_1} - \dot{\theta}_0\big|_{t=t_0} \\
\Delta_a &= (t - t_0)/(t_1 - t_0) \\
\Delta_d &= (t - t_1)/(t_2 - t_1)
\end{aligned}
\tag{3.52}
$$

Here, t_0 is the initial time of start of trunk body motion, t_1 is the time taken to reach the maximum angular displacement rate, t_2 is the time of start of retardation of the trunk body, t_3 is the total time of motion of the robot. Once the angular constraint derivative $\mathbf{g}(t)$ is calculated, the angular velocity of the trunk body ($^G\boldsymbol{\omega}_0$) can be calculated from the expression as mentioned below.

$$
^G\boldsymbol{\omega}_0 = \begin{pmatrix} \cos \beta_0 \cos \theta_0 \dot{\alpha}_0 + \sin \theta_0 \dot{\beta}_0 \\ -\cos \beta_0 \sin \theta_0 \dot{\alpha}_0 + \cos \theta_0 \dot{\beta}_0 \\ \sin \beta_0 \dot{\alpha}_0 + \dot{\theta}_0 \end{pmatrix}
\tag{3.53}
$$

where $\dot{\boldsymbol{\eta}}_0 = \begin{bmatrix} \dot{\alpha}_0 & \dot{\beta}_0 & \dot{\theta}_0 \end{bmatrix}^{\mathrm{T}}$ are the rate of change of angular displacement of the trunk body along their respective axes.

(b) *Calculation of translational displacement rate inputs*

Mathematically,
Translational displacement rate is given by

$$
\dot{\mathbf{r}}^G_{P_0 O} = \mathbf{f}(t) \quad \in \mathbb{R}^3,
\tag{3.54}
$$

where the function $\mathbf{f}(t)$ governs the translational motion of the trunk body at any instant of time.

Straight-forward Motion
During straight-forward motion, the translational displacement of the trunk body is considered along Y-axis with respect to G.

Along X-direction:

$$
f_x(t) = 0
\tag{3.55}
$$

Along Y-direction:

$$
\begin{aligned}
f_y(t) &= \dot{y}^G_{P_0O}\big|_{t=t_0} + a_{v_y}.\Delta_a^2.(3 - 2\Delta_a) \text{ for } t_0 \text{ to } t_1 \\
&= \dot{y}^G_{P_0O}\big|_{t=t_1} \qquad\qquad\qquad\qquad\quad \text{for } t_1 \text{ to } t_2 \\
&= \dot{y}^G_{P_0O}\big|_{t=t_1} - a_{v_y}.\Delta_d^2.(3 - 2\Delta_d) \quad \text{for } t_2 \text{ to } t_3
\end{aligned}
\tag{3.56}
$$

where,

$$
\begin{aligned}
\Delta_a &= (t - t_0)/(t_1 - t_0) \\
\Delta_d &= (t - t_2)/(t_3 - t_2) \\
a_{v_y} &= \dot{y}^G_{P_0O}\big|_{t=t_1} - \dot{y}^G_{P_0O}\big|_{t=t_0}
\end{aligned}
\tag{3.57}
$$

Here, $\dot{y}^G_{P_0O}\big|_{t=t_0}$ and $\dot{y}^G_{P_0O}\big|_{t=t_1}$ are translational displacement rates of the trunk body along Y-direction of frame $\mathbf{G}$ at time $t = t_0$ and $t = t_1$, respectively. Similarly,

Along Z-direction:
Cycle-1

$$
\begin{aligned}
f_z(t) &= \dot{z}^G_{P_0O}\big|_{t=t_0} + a_{v_z}\big|_1.\Delta_{as}\big|_1^2.(3 - 2\Delta_{as}\big|_1) \text{ for } t_0 \text{ to } t_1\big|_1 \\
&= \dot{z}^G_{P_0O}\big|_{t=t_1\big|_1} \qquad\qquad\qquad\qquad\qquad \text{for } t = t_1\big|_1 \\
&= \dot{z}^G_{P_0O}\big|_{t=t_1\big|_1} - a_{v_z}\big|_1.\Delta_{ds}\big|_1^2.(3 - 2\Delta_{ds}\big|_1) \text{ for } t_1\big|_1 \text{ to } t_3^s\big|_1
\end{aligned}
\tag{3.58}
$$

where

$$
\begin{aligned}
a_{v_z}\big|_1 &= \dot{z}^G_{P_0O}\big|_{t=t_1\big|_1} - \dot{z}^G_{P_0O}\big|_{t=t_0}; \\
\Delta_{as}\big|_1 &= (t - t_0)/(t_1\big|_1 - t_0); \\
\Delta_{ds}\big|_1 &= (t - t_1\big|_1)\big/(t_3^s\big|_1 - t_1\big|_1).
\end{aligned}
\tag{3.59}
$$

Here, $\dot{z}^G_{P_0O}\big|_{t=t_0}$ and $\dot{z}^G_{P_0O}\big|_{t=t_1\big|_1}$ are translational displacement rates of the trunk body along Z-direction at time $t = t_0$ and $t = t_1\big|_1$, respectively. Similarly,
Cycle-2

$$
\begin{aligned}
f_z(t) &= \dot{z}^G_{P_0O}\big|_{t=t_3^s\big|_1} - a_{v_z}\big|_2.\Delta_{as}\big|_2^2.(3 - 2\Delta_{as}\big|_2) \text{ for } t_3^s\big|_1 \text{ to } t_1\big|_2 \\
&= -\dot{z}^G_{P_0O}\big|_{t=t_1\big|_2} \qquad\qquad\qquad\qquad\qquad \text{for } t = t_1\big|_2 \\
&= -\dot{z}^G_{P_0O}\big|_{t=t_1\big|_2} + a_{v_z}\big|_2.\Delta_{ds}\big|_2^2.(3 - 2\Delta_{ds}\big|_2) \text{ for } t_1\big|_2 \text{ to } t_3^s\big|_2
\end{aligned}
\tag{3.60}
$$

where

$$
\begin{aligned}
a_{v_z}\big|_2 &= \dot{z}^G_{P_0O}\big|_{t=t_1\big|_2} - \dot{z}^G_{P_0O}\big|_{t=t_3^s\big|_1}; \\
\Delta_{as}\big|_2 &= (t - t_3^s\big|_1)\big/(t_1\big|_2 - t_3^s\big|_1); \\
\Delta_{ds}\big|_2 &= (t - t_1\big|_2)\big/(t_3^s\big|_2 - t_1\big|_2).
\end{aligned}
\tag{3.61}
$$

Here, $\dot{z}^G_{P_0O}\big|_{t=t^s_3|_1}$ and $\dot{z}^G_{P_0O}\big|_{t=t_1|_2}$ are translational displacement rates of the trunk body along Z-direction at time $t = t^s_3\big|_1$ and $t = t_1|_2$, respectively. Similarly, Cycle-3

$$
\begin{aligned}
f_z(t) &= \dot{z}^G_{P_0O}\big|_{t=t^s_3|_2} + a_{v_z}\big|_3 . \Delta_{as}\big|^2_3.(3 - 2\Delta_{as}|_3) \text{ for } t^s_3\big|_2 \text{ to } t_1|_3 \\
&= \dot{z}^G_{P_0O}\big|_{t=t_1|_3} \qquad\qquad\qquad\qquad\qquad\qquad \text{for } t = t_1|_3 \qquad\qquad (3.62) \\
&= \dot{z}^G_{P_0O}\big|_{t=t_1|_3} - a_{v_z}\big|_3 . \Delta_{ds}\big|^2_3.(3 - 2\Delta_{ds}|_3) \quad \text{for } t_1|_3 \text{ to } t_3
\end{aligned}
$$

where

$$
\begin{aligned}
a_{v_z}\big|_3 &= \dot{z}^G_{P_0O}\big|_{t=t_1|_3} - \dot{z}^G_{P_0O}\big|_{t=t^s_3|_2} ; \\
\Delta_{as}\big|_3 &= (t - t^s_3\big|_3)\big/(t_1|_3 - t^s_3\big|_3); \\
\Delta_{ds}\big|_3 &= (t - t_1|_3)\big/(t_3 - t_1|_3).
\end{aligned} \qquad (3.63)
$$

Here, $\dot{z}^G_{P_0O}\big|_{t=t^s_3|_2}$ and $\dot{z}^G_{P_0O}\big|_{t=t_1|_3}$ are translational displacement rates of the trunk body along Z-direction at time $t = t^s_3\big|_2$ and $t = t_1|_3$, respectively.

Crab Motion

Along X-direction:

During crab motion, the velocity function $f_x(t)$ along X-direction of frame G is given by:

$$
\begin{aligned}
f_x(t) &= \dot{x}^G_{P_0O}\big|_{t=t_0} + a_{v_x}.\Delta^2_a.(3 - 2\Delta_a) \text{ for } t_0 \text{ to } t_1 \\
&= \dot{x}^G_{P_0O}\big|_{t=t_1} \qquad\qquad\qquad\qquad \text{for } t_1 \text{ to } t_2 \qquad (3.64) \\
&= \dot{x}^G_{P_0O}\big|_{t=t_1} - a_{v_x}.\Delta^2_d.(3 - 2\Delta_d) \quad \text{for } t_2 \text{ to } t_3
\end{aligned}
$$

where,

$$
\begin{aligned}
\Delta_a &= (t - t_0)/(t_1 - t_0) \\
\Delta_d &= (t - t_2)/(t_3 - t_2) \\
a_{v_x} &= \dot{x}^G_{P_0O}\big|_{t=t_1} - \dot{x}^G_{P_0O}\big|_{t=t_0}
\end{aligned} \qquad (3.65)
$$

Here, $\dot{x}^G_{P_0O}\big|_{t=t_0}$ and $\dot{x}^G_{P_0O}\big|_{t=t_1}$ are translational displacement rates of the trunk body along X-direction at time $t = t_0$ and $t = t_1$, respectively. Similarly,

Along Y-direction:

$$
f_y(t) = f_x(t).\tan\theta_c \qquad (3.66)
$$

where θ_c is the crab angle.

Along Z-direction:
Cycle-1

$$f_z(t) = \dot{z}^G_{P_0O}\big|_{t=t_0} + a_{v_z}\big|_1 \cdot \Delta_{as}\big|_1^2 \cdot (3 - 2\,\Delta_{as}\big|_1) \text{ for } t_0 \text{ to } t_1|_1$$
$$= \dot{z}^G_{P_0O}\big|_{t=t_1|_1} \qquad\qquad\qquad \text{for } t = t_1|_1 \qquad\qquad (3.67)$$
$$= \dot{z}^G_{P_0O}\big|_{t=t_1|_1} - a_{v_z}\big|_1 \cdot \Delta_{ds}\big|_1^2 \cdot (3 - 2\,\Delta_{ds}\big|_1) \text{ for } t_1|_1 \text{ to } t_3^s\big|_1$$

where

$$a_{v_z}\big|_1 = \dot{z}^G_{P_0O}\big|_{t=t_1|_1} - \dot{z}^G_{P_0O}\big|_{t=t_0};$$
$$\Delta_{as}\big|_1 = (t - t_0)/(t_1|_1 - t_0);$$
$$\Delta_{ds}\big|_1 = (t - t_1|_1)/(t_3^s\big|_1 - t_1|_1). \qquad\qquad (3.68)$$

Here, $\dot{z}^G_{P_0O}\big|_{t=t_0}$ and $\dot{z}^G_{P_0O}\big|_{t=t_1|_1}$ are translational displacement rates of the trunk body along Z-direction at time $t = t_0$ and $t = t_1|_1$, respectively. Similarly,
Cycle-2

$$f_z(t) = \dot{z}^G_{P_0O}\big|_{t=t_3^s|_1} - a_{v_z}\big|_2 \cdot \Delta_{as}\big|_2^2 \cdot (3 - 2\,\Delta_{as}\big|_2) \text{ for } t_3^s\big|_1 \text{ to } t_1|_2$$
$$= -\dot{z}^G_{P_0O}\big|_{t=t_1|_2} \qquad\qquad\qquad \text{for } t = t_1|_2 \qquad\qquad (3.69)$$
$$= -\dot{z}^G_{P_0O}\big|_{t=t_1|_2} + a_{v_z}\big|_2 \cdot \Delta_{ds}\big|_2^2 \cdot (3 - 2\,\Delta_{ds}\big|_2) \text{ for } t_1|_2 \text{ to } t_3^s\big|_2$$

where

$$a_{v_z}\big|_2 = \dot{z}^G_{P_0O}\big|_{t=t_1|_2} - \dot{z}^G_{P_0O}\big|_{t=t_3^s|_1};$$
$$\Delta_{as}\big|_2 = (t - t_3^s\big|_1)/(t_1|_2 - t_3^s\big|_1);$$
$$\Delta_{ds}\big|_2 = (t - t_1|_2)/(t_3^s\big|_2 - t_1|_2). \qquad\qquad (3.70)$$

Here, $\dot{z}^G_{P_0O}\big|_{t=t_3^s|_1}$ and $\dot{z}^G_{P_0O}\big|_{t=t_1|_2}$ are translational displacement rates of the trunk body along Z-direction at time $t = t_0$ and $t = t_1|_2$, respectively.
Cycle-3

$$f_z(t) = \dot{z}^G_{P_0O}\big|_{t=t_3^s|_2} + a_{v_z}\big|_3 \cdot \Delta_{as}\big|_3^2 \cdot (3 - 2\,\Delta_{as}\big|_3) \text{ for } t_3^s\big|_2 \text{ to } t_1|_3$$
$$= \dot{z}^G_{P_0O}\big|_{t=t_1|_3} \qquad\qquad\qquad \text{for } t = t_1|_3 \qquad\qquad (3.71)$$
$$= \dot{z}^G_{P_0O}\big|_{t=t_1|_3} - a_{v_z}\big|_3 \cdot \Delta_{ds}\big|_3^2 \cdot (3 - 2\,\Delta_{ds}\big|_3) \qquad \text{for } t_1|_3 \text{ to } t_3$$

where

$$a_{v_z}\big|_3 = \dot{z}^G_{P_0O}\big|_{t=t_1|_3} - \dot{z}^G_{P_0O}\big|_{t=t_3^s|_2};$$
$$\Delta_{as}\big|_3 = (t - t_3^s\big|_3)/(t_1|_3 - t_3^s\big|_3);$$

$$\Delta_{ds}|_3 = (t - t_1|_3)/(t_3 - t_1|_3). \tag{3.72}$$

Here, $\dot{z}^G_{P_0 O}\big|_{t=t^s_3|_2}$ and $\dot{z}^G_{P_0 O}\big|_{t=t_1|_3}$ are translational displacement rates of the trunk body along Z-direction at time $t = t^s_3|_2$ and $t = t_1|_3$, respectively.

Turning Motion

Along X-direction:

During turning motion, the velocity function $f_x(t)$ along X-direction of frame G is given by:

$$f_x(t) = -\rho_0 g_z(t) \sin \theta_0, \tag{3.73}$$

where ρ_0 is the turning radius of point P_0.

Along Y-direction:

$$f_y(t) = -\rho_0 g_z(t) \cos \theta_0, \tag{3.74}$$

Along Z-direction:
Cycle-1

$$
\begin{aligned}
f_z(t) &= \dot{z}^G_{P_0 O}\big|_{t=t_0} + a_{v_z}\big|_1 \cdot \Delta_{as}|^2_1 \cdot (3 - 2\Delta_{as}|_1) \text{ for } t_0 \text{ to } t_1|_1 \\
&= \dot{z}^G_{P_0 O}\big|_{t=t_1|_1} \qquad\qquad\qquad\quad \text{ for } t = t_1|_1 \\
&= \dot{z}^G_{P_0 O}\big|_{t=t_1|_1} - a_{v_z}\big|_1 \cdot \Delta_{ds}|^2_1 \cdot (3 - 2\Delta_{ds}|_1) \text{ for } t_1|_1 \text{ to } t^s_3|_2
\end{aligned}
\tag{3.75}
$$

where

$$
\begin{aligned}
a_{v_z}\big|_1 &= \dot{z}^G_{P_0 O}\big|_{t=t_1|_1} - \dot{z}^G_{P_0 O}\big|_{t=t_0}; \\
\Delta_{as}|_1 &= (t - t_0)/(t_1|_1 - t_0); \\
\Delta_{ds}|_1 &= (t - t_1|_1)/(t^s_3\big|_1 - t_1|_1).
\end{aligned}
\tag{3.76}
$$

Here, $\dot{z}^G_{P_0 O}\big|_{t=t_0}$ and $\dot{z}^G_{P_0 O}\big|_{t=t_1|_1}$ are translational displacement rates of the trunk body along Z-direction at time $t = t_0$ and $t = t_1|_1$, respectively. Similarly, Cycle-2

$$
\begin{aligned}
f_z(t) &= \dot{z}^G_{P_0 O}\big|_{t=t^s_3|_1} - a_{v_z}\big|_2 \cdot \Delta_{as}|^2_2 \cdot (3 - 2\Delta_{as}|_2) \text{ for } t^s_3\big|_1 \text{ to } t_1|_2 \\
&= -\dot{z}^G_{P_0 O}\big|_{t=t_1|_2} \qquad\qquad\qquad\quad \text{ for } t = t_1|_2 \\
&= -\dot{z}^G_{P_0 O}\big|_{t=t_1|_2} + a_{v_z}\big|_2 \cdot \Delta_{ds}|^2_2 \cdot (3 - 2\Delta_{ds}|_2) \text{ for } t_1|_2 \text{ to } t^s_3|_2
\end{aligned}
\tag{3.77}
$$

where

$$
\begin{aligned}
a_{v_z}\big|_2 &= \dot{z}^G_{P_0 O}\big|_{t=t_1|_2} - \dot{z}^G_{P_0 O}\big|_{t=t^s_3|_1}; \\
\Delta_{as}|_2 &= (t - t^s_3\big|_1)/(t_1|_2 - t^s_3\big|_1); \\
\Delta_{ds}|_2 &= (t - t_1|_2)/(t^s_3\big|_2 - t_1|_2).
\end{aligned}
\tag{3.78}
$$

Here, $\dot{z}^G_{P_0O}\big|_{t=t^s_3|_1}$ and $\dot{z}^G_{P_0O}\big|_{t=t_1|_2}$ are translational displacement rates of the trunk body along Z-direction at time $t = t^s_3\big|_1$ and $t = t_1|_2$, respectively. Similarly,

Cycle-3

$$
\begin{aligned}
f_z(t) &= \dot{z}^G_{P_0O}\big|_{t=t^s_3|_2} + a_{v_z}\big|_3 \cdot \Delta_{as}\big|^2_3 \cdot (3 - 2\,\Delta_{as}|_3) \quad \text{for } t^s_3\big|_2 \text{ to } t_1|_3 \\
&= \dot{z}^G_{P_0O}\big|_{t=t_1|_3} \qquad\qquad\qquad\qquad\qquad\qquad\quad \text{for } t = t_1|_3 \qquad\qquad (3.79) \\
&= \dot{z}^G_{P_0O}\big|_{t=t_1|_3} - a_{v_z}\big|_3 \cdot \Delta_{ds}\big|^2_3 \cdot (3 - 2\,\Delta_{ds}|_3) \quad \text{for } t_1|_3 \text{ to } t_3
\end{aligned}
$$

where

$$
\begin{aligned}
a_{v_z}\big|_3 &= \dot{z}^G_{P_0O}\big|_{t=t_1|_3} - \dot{z}^G_{P_0O}\big|_{t=t^s_3|_2}; \\
\Delta_{as}\big|_3 &= (t - t^s_3\big|_3)/(t_1|_3 - t^s_3\big|_3); \\
\Delta_{ds}\big|_3 &= (t - t_1|_3)/(t_3 - t_1|_3). \qquad\qquad\qquad\qquad (3.80)
\end{aligned}
$$

Here, $\dot{z}^G_{P_0O}\big|_{t=t^s_3|_2}$ and $\dot{z}^G_{P_0O}\big|_{t=t_1|_3}$ are translational displacement rates of the trunk body along Z-direction at time $t = t^s_3\big|_2$ and $t = t_1|_3$, respectively.

It is to be noted to solve the inverse kinematics of the system, the velocity of the trunk body, $\mathbf{v}^G_0 = (\dot{\mathbf{r}}^G_{P_0O}, {}^G\boldsymbol{\omega}_0) \in \mathbb{R}^6$, is provided as an important input in addition to others as discussed in the next subsections. Next, using the transformation matrix $\mathbf{A}^{G_0G}$, the velocity vector of the trunk body is transformed to frame $\boldsymbol{G}_0$.

3.2.5.2   Swing Leg Trajectory Planning

The swing leg trajectory planning of six-legged robots in 3D Cartesian space is an important consideration for locomotion on rough terrain and is one of the fundamental steps in the study of kinematics and dynamics of the system. A variety of swing leg trajectories for execution of straight-forward, crab, and turning motions have been proposed to achieve smooth motions on different terrains. However, it is to be noted that with straight-forward, crab or turning gait strategy, the trajectory planning is computationally intensive.

The robot in the current work has the capability to regulate its parameters in accordance with the terrain topology. The feet tip of the swing leg traverses through a predefined trajectory in 3D Cartesian space with the strokes of the swing legs placed on the outer side comparatively longer than those placed on the inner side. Now, the computation of the coordinates of the trajectory of the swing is carried out with respect to local frame $\boldsymbol{L}_{i3}$. Thereafter, computed data points (local coordinates) of the swing trajectory of the leg are converted to global coordinates with respect to the frame $\boldsymbol{G}$.

The steps mentioned below are essential for the calculation of the tip-point trajectory coordinates during swing phase and feet-hold position of the leg during stance phase.

Step 1. Select initial inputs to calculate the trajectory of swing leg i (with respect to frame G). The relevant inputs are as follows:

For trunk body	(a) Linear stroke, s_0 (for straight-forward or crab motion) or angular stroke s_0^c (for turning motion), (b) initial position and orientation, (c) maximum rate of change of displacement (both angular and translation).
For Legs	(a) initial joint positions with respect to local frames, (b) initial joint angles, (c) trajectory ascend (γ_{xz}, γ_{yz}) and descend angles $\left(\gamma'_{xz}, \gamma'_{yz}\right)$.
For Terrain	(a) terrain height at the initial point of support, i.e., h_{in} (height of point P_{i3} w.r.t. frame G) and subsequent heights at the end of swing phases, i.e., h'_{in}, (b) maximum height of terrain Hm_{in}, (c) minimum gap between maximum height of swing and maximum height of terrain Δh (refer to Fig. 3.4).
Other Relevant Inputs	(a) crab angle, θ_c, and (b) turning radius, ρ_0.

Step 2. Calculation of the foot tip's motion parameters during swing leg trajectory planning with respect to the frame L_{i3} (refer to Fig. 3.4) using BCs P_{i3} (initial foothold position), Q_{i3} (point at which swing height reaches the maximum and translational velocity along Z with respect to frame G is zero), R_{i3} (point at which translational velocity along X with respect to frame G reaches the maximum), T_{i3} (point of start of retardation), and P'_{i3} (point of foothold after the end of swing phase). Also, the turning foot-tip radius ρ_i, as shown in Fig. 3.4, is calculated from the forward kinematics discussed in Sect. 3.2.2. For more details of trajectory planning of swing leg, readers may refer to Appendix A.4.

Step 3. Transformation of the foot tip's motion parameters from local reference frame L_{i3} to Quadrant Infotech (I) Pv frame G.

It is necessary to calculate the foot tip's motions parameters (position, velocity, acceleration, etc.) with respect to frame G once the same has been done with respect to L_{i3}. Vector loop equations with suitable transformation matrices as stated below (refer to Fig. 3.4) are used. Therefore,

$$\mathbf{r}^G_{P^s_{i3}O} = \mathbf{r}^G_{P_{i3}O} + \mathbf{A}^{GL_{i3}}.\mathbf{r}^{L_{i3}}_{P^s_{i3}P_{i3}} \quad \text{(Straight-forward)} \tag{3.81}$$

$$\mathbf{r}^G_{P^s_{i3}O} = \mathbf{r}^G_{P_{i3}O} + \mathbf{A}^{GL''_{i3}}.\mathbf{r}^{L''_{i3}}_{P^s_{i3}P_{i3}} \quad \text{(Crab)} \tag{3.82}$$

$$\mathbf{r}^G_{P^s_{i3}O} = \mathbf{r}^G_{P_{i3}O} + \mathbf{A}^{GL''_{i3}}.\mathbf{r}^{L''_{i3}}_{P^s_{i3}P_{i3}} \quad \text{(Turning)} \tag{3.83}$$

Here, the suffix s for the point P^s_{i3} represents swing, $\mathbf{A}^{GL_{i3}} = \mathbf{I}_3$ (identity matrix), $\mathbf{A}^{GL''_{i3}} = \mathbf{A}^{L_{i3}L''_{i3}}$, $\mathbf{A}^{GL'''_{i3}} = \mathbf{A}^{L_{i3}L'''_{i3}}$. One may refer to Appendix A.4 for more details.

Step 4.   Transformation of the foot tip's motion parameters from frame G to G_0 for a fixed orientation vector $\boldsymbol{\eta}_G$ using the transform $\mathbf{A}^{G_0 G}$.

It is to be noted that the θ_c shown in Fig. 3.4b is kept fixed for a given gait cycle, whereas $\theta_{i3}^{L_{i3}}$ shown in Fig. 3.4c varies with respect to time for a given gait cycle. The absolute range of θ_c for the present robotic system is limited to $20° \leq \theta_c \leq 90°$. Moreover, in this book, it is assumed that the full swing stroke of swing leg i (i.e., s_w^f) is twice the body stroke s_0 (linear) or s_0^c (angular) for straight-forward and turning motion while for crab motion, $s_w^f = s_0$.

3.2.5.3   Foot Slip During Support Phase

To make the motion of the legs more accurate, an attempt has been made by consideration foot slip during foot–terrain interaction in support phase [2]. The translational velocity along Z-direction approaches to zero gradually and subsequently, slip occurs till the beginning of the next swing phase. It is assumed that the slip of the leg tip takes place in XY plane of reference frame G, at a slip angle, ε_s and the slip velocity, $v_{xy}^{N_{i3}}$ with respect to local frame N_{i3} (refer to Fig. 3.6a) instantaneously becomes equal to zero at the time of lift from point $\mathbf{P}_{i3}$ of the next phase. Further, the slip velocity of the support leg instantaneously approaches zero at the time of takeoff at point $\mathbf{P}_{i3}$. Also, the slip trajectory, on simplification, is a straight path, which initiates at the beginning of support phase (point $\mathbf{P}'_{i3}$) and terminates at the start of next swing phase $\mathbf{P}_{i3}$ (refer to Fig. 3.6a). The slip velocity $v_{xy}^{N_{i3}}$ is assumed to be regulated by a cubic polynomial like Eq. (3.41) and is given by

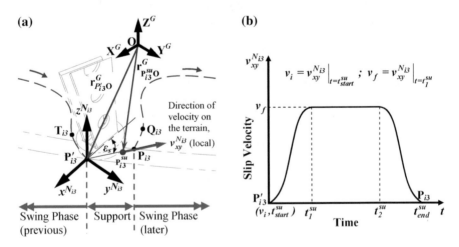

Fig. 3.6 Foot slip during support phase of ith leg **a** a scheme of foot-terrain interaction kinematics, **b** slip velocity versus time

$$v_{xy}^{N_{i3}}(t) = v_{xy}^{N_{i3}}\big|_{t=t_0^{su}} + a_{xy}.\Delta_{a_{xy}}^2 (3 - 2\Delta_{a_{xy}}) \quad \text{for } t_{start}^{su} \text{ to } t_1^{su}$$

$$= v_{xy}^{N_{i3}}\big|_{t=t_1^{su}} \qquad\qquad\qquad \text{for } t_1^{su} \text{ to } t_2^{su} \qquad (3.84)$$

$$= v_{xy}^{N_{i3}}\big|_{t=t_2^{su}} - a_{xy}.\Delta_{d_{xy}}^2 (3 - 2\Delta_{d_{xy}}), \quad \text{for } t_2^{su} \text{ to } t_{end}^{su}$$

where for every support phase of leg i, t_{start}^{su}, t_1^{su}, t_2^{su}, and t_{end}^{su} (suffix su represents support) indicate the initial start time (at point $\mathbf{P}_{i3}'$), time taken to reach maximum slip velocity, time of start of retardation, and end time of leg's slip (at point $\mathbf{P}_{i3}$ of the next swing phase), respectively, (refer to Fig. 3.6b). Also,

$$a_{xy} = v_{xy}^{N_{i3}}\big|_{t=t_1^{su}} - v_{xy}^{N_{i3}}\big|_{t=t_0^{su}} \qquad (3.85)$$

$$\Delta_{a_{xy}} = \left(t - t_0^{su}\right)/\left(t_1^{su} - t_0^{su}\right) \qquad (3.86)$$

$$\Delta_{d_{xy}} = \left(t - t_1^{su}\right)/\left(t_2^{su} - t_1^{su}\right) \qquad (3.87)$$

The values of t_{start}^{su} and t_{end}^{su} can easily be computed for a leg in support phase during a duty cycle as discussed in the next Sect. 3.2.6. Further, assuming the relation as mentioned below, the values of t_1^{su} and t_2^{su} can also be obtained.

$$\Delta t^{su} = t_1^{su} - t_{start}^{su} = t_{end}^{su} - t_2^{su} \qquad (3.88)$$

The slip velocity with respect to reference frame $\mathbf{G}$ is given by the expression:

$$v_{xy}^G = \mathbf{A}^{GN_{i3}}.v_{xy}^{N_{i3}} \qquad (3.89)$$

Here, the transformation matrix $\mathbf{A}^{GN_{i3}} = \mathbf{I}_3$, since the frame local N_{i3} is assumed to be parallel to frame $\mathbf{G}$.

The slip velocities components along horizontal and vertical axes are given by the subsequent expressions:

$$v_x^G = v_{xy}^G.\cos \varepsilon_s = v_{xy}^{N_{i3}}.\cos \varepsilon_s \qquad (3.90)$$

$$v_y^G = v_{xy}^G.\sin \varepsilon_s = v_{xy}^{N_{i3}}.\sin \varepsilon_s \qquad (3.91)$$

The displacement of the leg tip with respect to frame $\mathbf{G}$ at any instant of time during slip (i.e., $\mathbf{r}_{P_{i3}^{su}O}^G$ as shown in Fig. 3.6a) can be computed by integrating Eqs. (3.90) and (3.91) with respect to time t. Thereafter, using the transformation matrix $\mathbf{A}^{G_0G}$, the displacement of the tip can be computed with respect to static global reference frame $\mathbf{G}_0$. Here, it is assumed that the magnitude of slip velocity is small. If there is too much slip of the leg tip, the robotic structure will be staggered and the preferred motion of the robot for achieving a specific task may not be possible. It is to be noted that the importance of foot–terrain interaction mechanics lies in its dealing with coupled-dynamics problems as discussed in Chap. 4.

3.2.6 Gait Planning Strategy

For the motion of the robot's legs to be in sequential manner, it is mandatory to have an efficient algorithm and an effective gait planning to focus on the movement of the trunk body and legs walking on the straight, curved, or crab path in various terrains. In this book, the gait strategies with duty factor (DF) = 1/2, 2/3, 3/4 (refer to Fig. 3.7) for locomotion have been emphasized accompanied by the total gait cycle time, swing phase time, and stance phase time [5]. Throughout the straight-forward and crab motions, the robot moves in a straight-forward or crab path with linear stroke, s_0 of the trunk body, whereas during turning motion, the robot starts to move in a circular path with radius, ρ_0 (at point $\mathbf{P}_0$) and angular stroke, s_0^c of the trunk body.

To generalize the calculation for motion, initially the rate of change of displacement (linear or angular) is taken as v_{t_0} for the trunk body with respect to point $\mathbf{P}_0$ is considered to be, and while the maximum rate of change of displacement (linear or angular) is v_{t_1}. Figure 3.7 shows the rate of change of displacement of the trunk

Fig. 3.7 Wave gaits under investigation **a** DF $= 1/2$, **b** DF $= 2/3$, **c** DF $= 3/4$

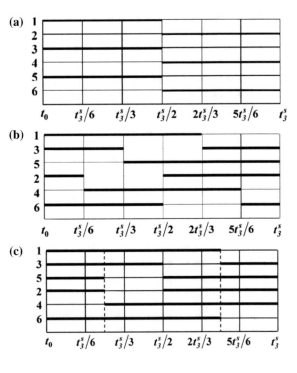

body with respect to frame G with n gait cycles (a) along Y-axis for straight-forward motion, (b) along X-axis for crab motion, and (c) along Z-axis for turning motion.

For the robot, the total time of motion is equivalent to the time taken to complete n duty cycles, i.e., $t_3^s\big|_n = t_3$ (where suffix s represents swing) and is calculated as follows:

$$t_3 = t_0 + 2\Delta t + (1/v_{t_1})[s_T - (v_{t_0} + v_{t_0 t_1}/2).(1/\Delta_a') - (v_{t_1} - v_{t_0 t_1}/2).(1/\Delta_d')] \tag{3.92}$$

$$v_{t_0 t_1} = v_{t_0} - v_{t_1}, \quad (\text{carb}) \tag{3.93}$$

$$\Delta t = t_1 - t_0 = t_3 - t_2, (\text{assumed}) \tag{3.94}$$

$$s_T = n.(m s_0''), \tag{3.95}$$

where s_T is the total distance covered; m represents the number of divisions of a gait cycle (in this case: $m = 6$); Figure 3.8 shows the displacement of trunk body s_0'' per division in a gait cycle; Δt is the duration of acceleration and deceleration of the trunk body; Δ_a' and Δ_d' denote the first-order derivatives of Δ_a and Δ_d with respect to time, respectively; t_0, t_1, and t_2 have usual meanings as discussed in Sect. 3.2.5.1. Refer to Appendix A.5 for the detailed time calculations in gait planning.

In the present work,

$$v_{t_0} = \dot{y}_{P_0 O}^G\big|_{t=t_0} (\text{straight}-\text{forward}) \tag{3.96}$$

$$= \dot{x}_{P_0 O}^G\big|_{t=t_0} (\text{crab}) \tag{3.97}$$

$$= \rho_0 \dot{\theta}_0\big|_{t=t_0} (\text{turning}) \tag{3.98}$$

and

$$v_{t_1} = \dot{y}_{P_0 O}^G\big|_{t=t_1} (\text{straight}-\text{forward}) \tag{3.99}$$

$$= \dot{x}_{P_0 O}^G\big|_{t=t_1} (\text{crab}) \tag{3.100}$$

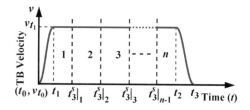

Fig. 3.8 Rate of change of displacement (linear or angular motion with respect to frame G) for the trunk body

$$= \rho_0 \dot{\theta}_0 \big|_{t=t_1} \text{(turning)} \qquad\qquad (3.101)$$

The termination time for each of the gait cycles is calculated as follows:

$$t_3^s \big|_1 = t_1 + \left(1/v_{t_1}\right)\left[m.s_0'' - \left(v_{t_0} + v_{t_0 t_1}/2\right).\left(1/\Delta_a'\right)\right] \qquad (3.102)$$

$$t_3^s \big|_2 = t_3^s \big|_1 + m.s_0''/v_{t_1} \qquad\qquad (3.103)$$

$$\vdots \quad \vdots \quad \vdots$$

$$t_3^s \big|_{n-1} = t_3^s \big|_{n-2} + m.s_0''/v_{t_1} \qquad\qquad (3.104)$$

$$t_3^s \big|_n = t_3, \qquad\qquad (3.105)$$

where $t_3^s \big|_1$, $t_3^s \big|_2$, ..., $t_3^s \big|_{n-1}$ and $t_3^s \big|_n$ are the termination time of gait Cycle-1, 2, ..., $(n-1)$ and n, respectively.

After calculating the end time for each gait cycle, the time of swing and support phase for each leg, during the cycle, are in accordance with the gait diagram, as shown in Fig. 3.8. This helps in developing a proper synchronization between the legs and trunk body.

3.2.7 Evaluation of Kinematic Parameters

The evaluation of joint angles with respect to time leads to that of the kinematic motion parameters for a specified gait and motion planning of the robot on various terrains. These parameters include both velocity (both translational and angular), acceleration (both translational and angular) of the links as well as trace of aggregate center of mass of the robot, etc. as discussed in Sect. 3.2.8. The velocity and acceleration vectors of the links $i1$, $i2$ and $i3$ of ith leg with respect to G are represented by, $\mathbf{v}_{ij}^G = ((\dot{\mathbf{r}}_{P_{ij}O}^G)^T, (^G\boldsymbol{\omega}_{ij})^T)^T \in \mathbb{R}^{108}$, $\dot{\mathbf{v}}_{ij}^G = ((\ddot{\mathbf{r}}_{P_{ij}O}^G)^T, (^G\dot{\boldsymbol{\omega}}_{ij})^T)^T \in \mathbb{R}^{108}$, respectively, where $i = 1$ to 6, $j = 1$ to 3.

The translational velocities and accelerations of the links are calculated based on the constrained kinematic equations set represented by Eq. (3.23). On keen analysis of the locomotion of a legged robot, it is revealed that the calculation of angular velocities $(^{G_0}\boldsymbol{\omega}_{ij})$ and accelerations $(^{G_0}\dot{\boldsymbol{\omega}}_{ij})$ of the joints of all the legs with respect to global reference frame are very critical. They are dependent on the coordinates of both $\mathbf{P}_0$ and $\mathbf{P}_{i3}$, during the swing phase, whereas they are dependent only on the coordinates of $\mathbf{P}_0$, during the support phase. A summary of the mathematical details is as given below (refer to Appendix A.6 for details).

A. **For support legs,**

(i) *Link angular velocities*

Angular velocity vector of the link ij during support phase is given by

$$^G\boldsymbol{\omega}_{ij} = \mathbf{J}_{r_{ij}} \cdot \dot{\mathbf{p}}_0^G, \tag{3.106}$$

where $\mathbf{J}_{r_{ij}}$ is a Jacobian matrix, the details of which are given in Appendix A.7; $\dot{\mathbf{p}}_0^G$ is the first-order derivative of $\mathbf{p}_0^G$ (vector of cartesian coordinates) with respect to time [refer to (Eq. 3.1)].

(ii) *Link translational velocities*

For the links $i1$, $i2$, and $i3$ of leg i, the translational velocity vectors are given by $\dot{\mathbf{r}}_{P_{i1}O}^G, \dot{\mathbf{r}}_{P_{i2}O}^G$ and $\dot{\mathbf{r}}_{P_{i3}O}^G$, respectively. They are obtained by differentiating the functions, $\mathbf{g}_{3i}(\mathbf{p}_i^G)$, $\mathbf{g}_{5i}(\mathbf{p}_i^G)$, and $\mathbf{g}_{7i}(\mathbf{p}_i^G)$, respectively, of Equation set (3.23). Therefore,

$$\dot{\mathbf{r}}_{P_{i1}O}^G = \dot{\mathbf{r}}_{P_0O}^G + \mathbf{A}^{GL_0} \cdot {}^G\tilde{\boldsymbol{\omega}}_0 \cdot \mathbf{r}_{S_iP_0}^{L_0} + \mathbf{A}^{GL_i''} \cdot {}^G\tilde{\boldsymbol{\omega}}_{i1} \cdot \mathbf{r}_{P_{i1}S_i}^{L_i''}$$
$$= \dot{\mathbf{r}}_{P_0O}^G - \mathbf{A}^{GL_0} \cdot \tilde{\mathbf{r}}_{S_iP_0}^{L_0} \cdot {}^G\boldsymbol{\omega}_0 - \mathbf{A}^{GL_i''} \cdot \tilde{\mathbf{r}}_{P_{i1}S_i}^{L_i''} \cdot {}^G\boldsymbol{\omega}_{i1} \tag{3.107}$$

$$\dot{\mathbf{r}}_{P_{i2}O}^G = \dot{\mathbf{r}}_{P_{i1}O}^G + \mathbf{A}^{GL_{i1}''} \cdot {}^G\tilde{\boldsymbol{\omega}}_{i2} \cdot \mathbf{r}_{P_{i2}P_{i1}}^{L_{i1}''} = \dot{\mathbf{r}}_{P_{i1}O}^G - \mathbf{A}^{GL_{i1}''} \cdot \tilde{\mathbf{r}}_{P_{i2}P_{i1}}^{L_{i1}''} \cdot {}^G\boldsymbol{\omega}_{i2} \tag{3.108}$$

$$\dot{\mathbf{r}}_{P_{i3}O}^G = \dot{\mathbf{r}}_{P_{i2}O}^G + \mathbf{A}^{GL_{i2}''} \cdot {}^G\tilde{\boldsymbol{\omega}}_{i3} \cdot \mathbf{r}_{P_{i3}P_{i2}}^{L_{i2}''} = \dot{\mathbf{r}}_{P_{i2}O}^G - \mathbf{A}^{GL_{i2}''} \cdot \tilde{\mathbf{r}}_{P_{i3}P_{i2}}^{L_{i2}''} \cdot {}^G\boldsymbol{\omega}_{i3} \tag{3.109}$$

Here, $\mathbf{A}^{GL_0}, \mathbf{A}^{GL_i''}, \mathbf{A}^{GL_{i1}''}, \mathbf{A}^{GL_{i2}''}$ are the transformation matrices related to different local frames with respect to frame $\mathbf{G}$ (refer to Appendix A.3); $\mathbf{r}_{S_iP_0}^{L_0}$, $\mathbf{r}_{P_{i1}S_i}^{L_i''}$, $\mathbf{r}_{P_{i2}P_{i1}}^{L_{i1}''}$, $\mathbf{r}_{P_{i3}P_{i2}}^{L_{i2}''}$ are the local coordinates, respectively, obtained from Table 3.3; $^G\boldsymbol{\omega}_0$ is the trunk body angular velocity obtained using Eq. 3.53; $^G\boldsymbol{\omega}_{i1}$, $^G\boldsymbol{\omega}_{i2}$, and $^G\boldsymbol{\omega}_{i3}$ are the angular velocity vectors of the links $i1$, $i2$, and $i3$, respectively, and calculated using Eq. (3.106).

(iii) *Link angular accelerations*

The angular acceleration vector of the link ij during support phase is given by

$$^G\dot{\boldsymbol{\omega}}_{ij} = \mathbf{J}_{r_{ij}} \cdot \ddot{\mathbf{p}}_0^G + \dot{\mathbf{J}}_{r_{ij}} \cdot \dot{\mathbf{p}}_0^G, \tag{3.110}$$

where $\dot{\mathbf{J}}_{r_{ij}}$ is derivative of the Jacobian $\mathbf{J}_{r_{ij}}$, the details of which are given in Appendix A.7; $\ddot{\mathbf{p}}_0^G$ is the second-order derivative of $\mathbf{p}_0^G$ (vector of Cartesian coordinates).

(iv) *Link translational accelerations*

For the links $i1$, $i2$, and $i3$ of leg i, the translational acceleration vectors are given by, $\ddot{\mathbf{r}}_{P_{i1}O}^{G}$, $\ddot{\mathbf{r}}_{P_{i2}O}^{G}$, and $\ddot{\mathbf{r}}_{P_{i3}O}^{G}$, respectively. They are obtained by differentiating Eqs. (3.107), (3.108), and (3.109), respectively, with respect to time. Therefore,

$$\ddot{\mathbf{r}}_{P_{i1}O}^{G} = \ddot{\mathbf{r}}_{P_{0}O}^{G} - \mathbf{A}^{GL_{0}}.\tilde{\mathbf{r}}_{S_{i}P_{0}}^{L_{0}}.{}^{G}\dot{\boldsymbol{\omega}}_{0} - \mathbf{A}^{GL_{i}''}.\tilde{\mathbf{r}}_{P_{i1}S_{i}}^{L_{i}''}.{}^{G}\dot{\boldsymbol{\omega}}_{i1} + \mathbf{A}^{GL_{0}}.{}^{G}\tilde{\boldsymbol{\omega}}_{0}.{}^{G}\tilde{\boldsymbol{\omega}}_{0}.\mathbf{r}_{S_{i}P_{0}}^{L_{0}}$$
$$+ \mathbf{A}^{GL_{i}''}.{}^{G}\tilde{\boldsymbol{\omega}}_{i1}.{}^{G}\tilde{\boldsymbol{\omega}}_{i1}.\mathbf{r}_{P_{i1}S_{i}}^{L_{i}''} \tag{3.111}$$

$$\ddot{\mathbf{r}}_{P_{i2}O}^{G} = \ddot{\mathbf{r}}_{P_{i1}O}^{G} - \mathbf{A}^{GL_{i1}''}.\tilde{\mathbf{r}}_{P_{i2}P_{i1}}^{L_{i1}''}.{}^{G}\dot{\boldsymbol{\omega}}_{i2} + \mathbf{A}^{GL_{i1}''}.{}^{G}\tilde{\boldsymbol{\omega}}_{i2}.{}^{G}\tilde{\boldsymbol{\omega}}_{i2}.\mathbf{r}_{P_{i2}P_{i1}}^{L_{i1}''} \tag{3.112}$$

$$\ddot{\mathbf{r}}_{P_{i3}O}^{G} = \ddot{\mathbf{r}}_{P_{i2}O}^{G} - \mathbf{A}^{GL_{i2}''}.\tilde{\mathbf{r}}_{P_{i3}P_{i2}}^{L_{i2}''}.{}^{G}\dot{\boldsymbol{\omega}}_{i3} + \mathbf{A}^{GL_{i2}''}.{}^{G}\tilde{\boldsymbol{\omega}}_{i3}.{}^{G}\tilde{\boldsymbol{\omega}}_{i3}.\mathbf{r}_{P_{i3}P_{i2}}^{L_{i2}''} \tag{3.113}$$

Here, ${}^{G}\dot{\boldsymbol{\omega}}_{0}$ is the trunk body angular acceleration vector obtained by taking the derivative of ${}^{G}\boldsymbol{\omega}_{0}$ [refer to Eq. (3.53)], and for the links $i1$, $i2$, and $i3$, ${}^{G}\dot{\boldsymbol{\omega}}_{i1}$, ${}^{G}\dot{\boldsymbol{\omega}}_{i2}$ and ${}^{G}\dot{\boldsymbol{\omega}}_{i3}$ are the angular acceleration vectors, respectively, and calculated using Eq. (3.110).

B. For swing legs,

(i) *Link angular velocities*

During swing phase, the angular velocity vector of the link ij is given by

$$ {}^{G}\boldsymbol{\omega}_{ij} = \mathbf{J}_{r_{ij}}.\dot{\mathbf{p}}_{0}^{G} + \mathbf{J}_{r_{ij}}'.\dot{\mathbf{p}}_{i3}^{G}, \tag{3.114}$$

where $\mathbf{J}_{r_{ij}}$ and $\mathbf{J}_{r_{ij}}'$ represent the Jacobian matrices (detailed in Appendix A.7); $\dot{\mathbf{p}}_{0}^{G}$ and $\dot{\mathbf{p}}_{i3}^{G}$ are the first-order derivative of $\mathbf{p}_{0}^{G}$ and $\mathbf{p}_{i3}^{G}$, respectively, with respect to time [refer to Eqs. (3.1) and (3.2)].

(ii) *Link translational velocities*

For the links $i1$, $i2$ and $i3$ of leg i, the translational velocity vectors are obtained by differentiating the functions, $\mathbf{g}_{5i}(\mathbf{p}_{i}^{G})$, $\mathbf{g}_{7i}(\mathbf{p}_{i}^{G})$, and $\mathbf{g}_{9i}(\mathbf{p}_{i}^{G})$, respectively, of Equation set (3.23). Therefore,

$$\dot{\mathbf{r}}_{P_{i1}O}^{G} = \dot{\mathbf{r}}_{P_{i2}O}^{G} - \mathbf{A}^{GL_{i1}''}.{}^{G}\tilde{\boldsymbol{\omega}}_{i2}.\mathbf{r}_{P_{i2}P_{i1}}^{L_{i1}''} = \dot{\mathbf{r}}_{P_{i2}O}^{G} + \mathbf{A}^{GL_{i1}''}.\tilde{\mathbf{r}}_{P_{i2}P_{i1}}^{L_{i1}''}.{}^{G}\boldsymbol{\omega}_{i2} \tag{3.115}$$

$$\dot{\mathbf{r}}_{P_{i2}O}^{G} = \dot{\mathbf{r}}_{P_{i3}O}^{G} - \mathbf{A}^{GL_{i2}''}.{}^{G}\tilde{\boldsymbol{\omega}}_{i3}.\mathbf{r}_{P_{i3}P_{i2}}^{L_{i2}''} = \dot{\mathbf{r}}_{P_{i3}O}^{G} + \mathbf{A}^{GL_{i2}''}.\tilde{\mathbf{r}}_{P_{i3}P_{i2}}^{L_{i2}''}.{}^{G}\boldsymbol{\omega}_{i3} \tag{3.116}$$

$$\dot{\mathbf{r}}_{P_{i3}O}^{G} = \dot{\mathbf{r}}_{P_{i3}^{s}O}^{G} \tag{3.117}$$

Here, $\dot{\mathbf{r}}_{P_{i3}^{s}O}^{G}$ is obtained by taking time derivative of Eqs. (3.81), (3.82), and (3.83), respectively, for straight-forward, crab, and turning motion.

(iii) *Link angular accelerations*

The angular acceleration vector of the link ij during swing phase is given by,

$$^G\dot{\boldsymbol{\omega}}_{ij} = \mathbf{J}_{r_{ij}}\cdot\ddot{\mathbf{p}}_0^G + \dot{\mathbf{J}}_{r_{ij}}\cdot\dot{\mathbf{p}}_0^G + \mathbf{J}'_{r_{ij}}\cdot\ddot{\mathbf{p}}_{i3}^G + \dot{\mathbf{J}}'_{r_{ij}}\cdot\dot{\mathbf{p}}_{i3}^G, \tag{3.118}$$

where $\dot{\mathbf{J}}_{r_{ij}}$ and $\dot{\mathbf{J}}'_{r_{ij}}$ are derivatives of the Jacobian matrices $\mathbf{J}_{r_{ij}}$ and $\mathbf{J}'_{r_{ij}}$, respectively, the details of which are given in Appendix A.7; $\ddot{\mathbf{p}}_0^G$ and $\ddot{\mathbf{p}}_{i3}^G$ are the first-order derivatives of $\dot{\mathbf{p}}_0^G$ and $\dot{\mathbf{p}}_{i3}^G$, respectively, with respect to time [refer to Eqs. (3.1) and (3.2)].

(iv) *Link translational accelerations*

For the links $i1$, $i2$, and $i3$ of leg i, the translational acceleration vectors are obtained by differentiating Eqs. (3.115), (3.116), and (3.117), respectively, with respect to time. Therefore,

$$\ddot{\mathbf{r}}_{P_{i1}O}^G = \ddot{\mathbf{r}}_{P_{i2}O}^G + \mathbf{A}^{GL''_{i1}}.\tilde{\mathbf{r}}_{P_{i2}P_{i1}}^{L''_{i1}}.{}^G\dot{\boldsymbol{\omega}}_{i2} - \mathbf{A}^{GL''_{i1}}.{}^G\tilde{\boldsymbol{\omega}}_{i2}.{}^G\tilde{\boldsymbol{\omega}}_{i2}.\tilde{\mathbf{r}}_{P_{i2}P_{i1}}^{L''_{i1}} \tag{3.119}$$

$$\ddot{\mathbf{r}}_{P_{i2}O}^G = \ddot{\mathbf{r}}_{P_{i3}O}^G + \mathbf{A}^{GL''_{i2}}.\tilde{\mathbf{r}}_{P_{i3}P_{i2}}^{L''_{i2}}.{}^G\dot{\boldsymbol{\omega}}_{i3} - \mathbf{A}^{GL''_{i2}}.{}^G\tilde{\boldsymbol{\omega}}_{i3}.{}^G\tilde{\boldsymbol{\omega}}_{i3}.\tilde{\mathbf{r}}_{P_{i3}P_{i2}}^{L''_{i2}} \tag{3.120}$$

$$\ddot{\mathbf{r}}_{P_{i3}O}^G = \ddot{\mathbf{r}}_{P_{i3}^sO}^G \tag{3.121}$$

Here, $\ddot{\mathbf{r}}_{P_{i3}^sO}^G$ is obtained by taking time derivative of Eq. (3.117). $^G\dot{\boldsymbol{\omega}}_{i1}$, $^G\dot{\boldsymbol{\omega}}_{i2}$ and $^G\dot{\boldsymbol{\omega}}_{i3}$ are the angular acceleration vectors of the links $i1$, $i2$, and $i3$, respectively, and calculated using Eq. (3.118).

After calculating the kinematic motion parameters with respect to G, it is necessary to represent them with respect to global reference frame G_0. The translational velocities and accelerations are given by the following equations, respectively:

$$\dot{\mathbf{r}}_{P_{ij}O}^{G_0} = \mathbf{A}^{G_0G}.\dot{\mathbf{r}}_{P_{ij}O}^G, \tag{3.122}$$

$$\ddot{\mathbf{r}}_{P_{ij}O}^{G_0} = \mathbf{A}^{G_0G}.\ddot{\mathbf{r}}_{P_{ij}O}^G, \tag{3.123}$$

where $\mathbf{A}^{G_0G}$ is the transformation matrix for transformation of frame from G to G_0. Now, since in the present model the angular orientation vector $\boldsymbol{\eta}_G$ that defines the terrain topology (refer to Sect. 3.1) is constant, the angular motion parameters with respect to frame G_0 are equal to that with respect to frame G. Therefore,

$$^{G_0}\boldsymbol{\omega}_{ij} = {}^G\boldsymbol{\omega}_{ij} \tag{3.124}$$

$$^{G_0}\dot{\boldsymbol{\omega}}_{ij} = {}^G\dot{\boldsymbol{\omega}}_{ij} \tag{3.125}$$

Hence, with respect to $\boldsymbol{G}_0$, the velocity and acceleration vectors of the links $i1$, $i2$, and $i3$ of ith leg are represented by $\mathbf{v}_{ij}^{G_0} = ((\dot{\mathbf{r}}_{P_{ij}O}^{G_0})^{\mathrm{T}}, (^{G_0}\boldsymbol{\omega}_{ij})^{\mathrm{T}})^{\mathrm{T}} \in \mathbb{R}^{108}$, $\dot{\mathbf{v}}_{ij}^{G_0} = ((\ddot{\mathbf{r}}_{P_{ij}O}^{G_0})^{\mathrm{T}}, (^{G_0}\dot{\boldsymbol{\omega}}_{ij})^{\mathrm{T}})^{\mathrm{T}} \in \mathbb{R}^{108}$, respectively, where $i = 1$ to $6, j = 1$ to 3.

3.2.8 Estimation of Aggregate Center of Mass

The study of aggregate COM of the robotic system is crucial, if the robot is maneuvering over various terrain. For the stability of the robot, it is crucial to observe the COM of the system. It gives a proper perception of the leg characteristics, that is, whether the legs are able to provide necessary support to the trunk body of the system and there is a proper synchronization among the different legs. If the COM is not correctly scrutinized, a situation may arise, when the system may lose balance and falls, or there is a greater energy consumption due to inefficient motion. Mathematical calculations have been carried out to determine the variation of the aggregate COM of the system as mentioned below.

Position vector of the COMs of each component with respect to $\boldsymbol{G}$ (refer to Fig. 3.9),

$$\mathbf{r}_{C_0O}^G = \mathbf{r}_{P_0O}^G + \mathbf{r}_{C_0P_0}^G = \mathbf{r}_{P_0O}^G + \mathbf{A}^{GL_0}\mathbf{r}_{C_0P_0}^{L_0} \tag{3.126}$$

$$\mathbf{r}_{C_{i1}O}^G = \mathbf{r}_{P_{i1}O}^G + \mathbf{r}_{C_{i1}P_{i1}}^G = \mathbf{r}_{P_{i1}O}^G + \mathbf{A}^{GL_{i1}'}.\mathbf{r}_{C_{i1}P_{i1}}^{L_{i1}'} \tag{3.127}$$

$$\mathbf{r}_{C_{i2}O}^G = \mathbf{r}_{P_{i2}O}^G + \mathbf{r}_{C_{i2}P_{i2}}^G = \mathbf{r}_{P_{i2}O}^G + \mathbf{A}^{GL_{i2}'}.\mathbf{r}_{C_{i2}P_{i2}}^{L_{i2}'} \tag{3.128}$$

$$\mathbf{r}_{C_{i3}O}^G = \mathbf{r}_{P_{i3}O}^G + \mathbf{r}_{C_{i3}P_{i3}}^G = \mathbf{r}_{P_{i3}O}^G + \mathbf{A}^{GL_{i3}'}\mathbf{r}_{C_{i3}P_{i3}}^{L_{i3}'} \tag{3.129}$$

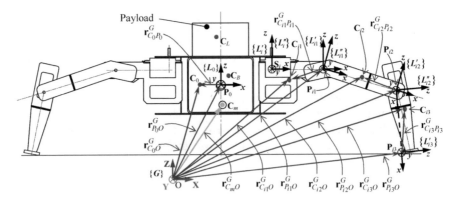

Fig. 3.9 Position of COM of each of the components of the robot

where

$$\mathbf{r}_{S_i O}^G = \mathbf{r}_{P_0 O}^G + \mathbf{A}^{GL_0} \mathbf{r}_{S_i P_0}^{L_0} \tag{3.130}$$

$$\mathbf{r}_{P_{i1} O}^G = \mathbf{r}_{S_i O}^G + \mathbf{A}^{GL_i''} \mathbf{r}_{P_{i1} S_i}^{L_i''} \tag{3.131}$$

$$\mathbf{r}_{P_{i2} O}^G = \mathbf{r}_{P_{i1} O}^G + \mathbf{A}^{GL_{i1}''} \mathbf{r}_{P_{i2} P_{i1}}^{L_{i1}''} \tag{3.132}$$

$$\mathbf{r}_{P_{i3} O}^G = \mathbf{r}_{P_{i2} O}^G + \mathbf{A}^{GL_{i2}''} \mathbf{r}_{P_{i3} P_{i2}}^{L_{i2}''} \tag{3.133}$$

Equations (3.130), (3.131), (3.132), and (3.133) are obtained from Table 3.2. It is to be taken into account that the parallel axis transformation leads to, $\mathbf{A}^{GL_{i1}'} = \mathbf{A}^{GL_i''} \mathbf{A}^{GL_{i2}'} = \mathbf{A}^{GL_{i1}''}$, and $\mathbf{A}^{GL_{i3}'} = \mathbf{A}^{GL_{i2}''}$.

The combined CM of trunk body (TB) and payload (PL) w.r.t frame L_0 can be written as,

$$\mathbf{r}_{C_0 P_0}^{L_0} = \frac{m_B \mathbf{r}_{C_B P_0}^{L_0} + m_L \mathbf{r}_{C_L P_0}^{L_0}}{m_0} \tag{3.134}$$

where

$$m_0 = m_B + m_L \tag{3.135}$$

Here, m_B, m_L, m_0 are the masses of the trunk body, payload and combined mass, respectively; $\mathbf{r}_{C_B P_0}^{L_0}, \mathbf{r}_{C_L P_0}^{L_0}$, and $\mathbf{r}_{C_0 P_0}^{L_0}$ are the displacement vector from point $\mathbf{P}_0$ to $\mathbf{C}_B, \mathbf{C}_L$ and $\mathbf{C}_0$, respectively, represented in reference frame $\mathbf{L}_0$.

The aggregate COM of the robotic system with respect to the frame $\mathbf{G}$ is given by:

$$\mathbf{r}_{C_m O}^G = \frac{m_0 \mathbf{r}_{C_0 O}^G + \sum_{i=1}^6 \sum_{j=1}^3 m_{ij} \mathbf{r}_{C_{ij} O}^G}{m_T} \tag{3.136}$$

where
Total mass of the system,

$$m_T = m_0 + \sum_{i=1}^6 \sum_{j=1}^3 m_{ij} \tag{3.137}$$

Here, m_{ij} is the mass of link l_{ij} ($i = 1$ to 6, $j = 1$ to 3), $\mathbf{r}_{C_0 O}^G$ and $\mathbf{r}_{C_{ij} O}^G$ are the displacement vectors from points $\mathbf{C}_0$ (location of combined center of mass (COM) of trunk body and payload) and $\mathbf{C}_{ij}$ (location of COM of link l_{ij}) represented in frame

Table 3.4 Coordinates of the local components of COM of the robotic system (in mm)

Component	Position vector		Local position coordinates (in mm)
Trunk body	$\mathbf{r}_{C_B P_0}^{L_0}$		$(0, 16, 74)^{\mathrm{T}}$
Payload	$\mathbf{r}_{C_L P_0}^{L_0}$		$(0, 72.5, 64.5)^{\mathrm{T}}$
Legs	$\mathbf{r}_{C_{ij} P_{ij}}^{L'_{ij}}$	$(i = 1, 3, 5; j = 1)$	$(71.217, -7.218, -14.036)^{\mathrm{T}}$
		$(i = 1, 3, 5; j = 2)$	$(71.443, -16.039, -0.183)^{\mathrm{T}}$
		$(i = 1, 3, 5; j = 3)$	$(99.253, -3.515, 0.059)^{\mathrm{T}}$
		$(i = 2, 4, 6; j = 1)$	$(-71.217, -7.218, -14.036)^{\mathrm{T}}$
		$(i = 2, 4, 6; j = 2)$	$(-71.443, -16.039, -0.183)^{\mathrm{T}}$
		$(i = 2, 4, 6; j = 3)$	$(-99.253, -3.515, 0.059)^{\mathrm{T}}$

G. The positional coordinates of the displacement vectors, $\mathbf{r}_{C_B P_0}^{L_0}$, $\mathbf{r}_{C_L P_0}^{L_0}$, and $\mathbf{r}_{C_{ij} P_{ij}}^{L'_{ij}}$ ($i = 1$ to $6; j = 1$ to 3) are given in Table 3.4.

Therefore, aggregate COM of the system with respect to frame $\mathbf{G}_0$ is given by

$$\mathbf{r}_{C_m O}^{G_0} = \mathbf{A}^{G_0 G} \cdot \mathbf{r}_{C_m O}^{G} \tag{3.138}$$

Equation (3.138) is generalized to calculate the aggregate center of mass of the robot locomotion over any kind of terrain with the predefined motion planning and gait planning.

3.3 Numerical Simulation: Study of Kinematic Motion Parameters

In this section, the capabilities of the developed kinematic model to tackle various terrain conditions are tested with three types of 3D examples, each with case studies adopting the wave gaits, since they are some of the standard gaits for six-legged walking robots. The input parameters, which can be controlled in the proposed model and the subsequent output parameters, which can be obtained from computer simulations are as listed below:

Simulation Inputs

- Initial position and orientation of the trunk body ($\mathbf{p}_0^G$)
- Initial joint angles ($\theta_{i1}, \beta_{i2}, \beta_{i3}$)
- Turning radius (ρ_0)
- Crab angle (θ_c)
- Translational displacement rate of the trunk body ($\dot{\mathbf{r}}_{P_0 O}^G\big|_{\max}$)
- Angular rate of the trunk body ($\dot{\boldsymbol{\eta}}_0\big|_{\max}$)
- Stroke of the trunk body (s_0/s_0^c)
- Duty factor (DF)

Fig. 3.10 Flowchart of computational algorithm for the inverse kinematic analysis

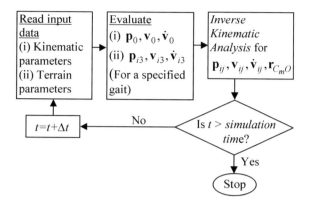

- Number of cycles (n)
- Terrain orientation ($\boldsymbol{\eta}_G$)
- h'_{in} [terrain height difference between nth foot-tip placement and $(n-1)$th foot-tip placement of swing leg i] measured with respect to frame $\boldsymbol{L}_{i3}$
- Maximum height of terrain in path of swing taken with respect to foot-tip placement at $(n-1)$th cycle (Hm_{in}) measured with respect to frame $\boldsymbol{L}_{i3}$
- Distance between leg tip and maximum height of terrain (Δh)
- Maximum slip velocity (v_f)
- Slip angle (ε_s)
- Lateral offset (l_i).

Simulation Outputs

- Joint position ($\mathbf{r}^{G_0}_{P_{ij}O}$ or $\boldsymbol{\eta}_{ij}$)
- Joint velocity ($\dot{\mathbf{r}}^{G_0}_{P_{ij}O}$ or $^{G_0}\boldsymbol{\omega}_{ij}$)
- Joint acceleration ($\ddot{\mathbf{r}}^{G_0}_{P_{ij}O}$ or $^{G_0}\dot{\boldsymbol{\omega}}_{ij}$)
- Position of the aggregate center of mass in meters ($\mathbf{r}^{G_0}_{C_mO}$).

To carry out the inverse kinematic analysis of the system, a flowchart is developed as given in Fig. 3.10. It describes the steps to be followed to carry out the analysis.

3.3.1 Case Study 1: Robot Motion in an Uneven Terrain with Straight-Forward Motion (DF = 1/2)

In the following case study, an effort has been made to carry out kinematic analysis of the realistic six-legged robot executing straight-forward motion on an uneven terrain. The required simulation input parameters are given as follows: The orientation and position of $\mathbf{P}_0$, at time $t = 0$, with respect to frame $\boldsymbol{G}$ are given by $\mathbf{p}^G_0 = \{0, 0.45, 0.15, -1, 2, 0\}^T$; Considering all the initial velocity components as zero, the maximum translational and angular displacement rates of the trunk body are given by $\dot{\mathbf{r}}^G_{P_0 O}\big|_{\max} = \{0, 0.15, 0.001\}^T$ and $\dot{\boldsymbol{\eta}}_0\big|_{\max} = \{-0.01, 0.02, 0\}^T$. Besides these, the other essential inputs are $\boldsymbol{\eta}_G = (0, 0, 0)^T$, $\theta_{i1} = \pm 20°$, $\beta_{i2} = \pm 16°$, $\beta_{i3} = \pm 69°$ (for $i = 1$ to 6). The other significant inputs are associated with the topography of the terrain, namely (i) Hm_{in} (maximum height of terrain in path of swing taken with respect to foot-tip placement at $(n-1)$th cycle) with respect to frame $\boldsymbol{L}_{i3}$ and (ii) h'_{in} (terrain height difference between nth foot-tip placement and $(n-1)$th foot-tip placement of swing leg i) with respect to frame $\boldsymbol{L}_{i3}$ (refer to Fig. 3.4). Hence, for $n = 3$, i.e., for three duty cycles, the values of Hm_{in} are in the order of $[Hm_{i1}, Hm_{i2}, Hm_{i3}]$ for $i = 1$ to 6, such that, $[0.06, 0.06, 0.06]$ corresponds to the respective heights of terrain, Hm_{in} the legs has to overcome during swing. Similarly, the values of h'_{in} are in the order of $[h'_{i1}, h'_{i2}, h'_{i3}]$ for $i = 1$ to 6, such that, $[0.005, 0.010, 0.005]$ corresponds to leg 1, $[0.010, 0.015, -0.005]$ corresponds to leg 2, $[0.005, 0.015, -0.01]$ corresponds to leg 3, $[0.02, 0.-015, -0.005]$ corresponds to leg 4, $[0.015, -0.005, 0.01]$ corresponds to leg 5 and $[0.02, -0.01, 0.005]$ corresponds to leg 6 (all values are in m). The value of Δh is kept constant, i.e., 0.002 m. Additionally, the maximum value of slip velocity is assumed to be zero with a slip angle of $\varepsilon_s = 0°$. The lateral offset (l_i) equal to 0.2185 m, is also kept constant. If not mentioned, the relevant units are as given in Sect. 3.3.

The simulations are run for three duty cycles, i.e., $n = 3$ and $s_0 = 0.15$ m with average time step $(h) = 0.05$ s in a commercially available compiler MATLAB. The total time taken to complete three cycles has been calculated as 6.95 s (first cycle: 2.45 s; second cycle: 2.05 s and third cycle: 2.45 s). The second cycle time is kept always less compare to first and third cycle due to the fact that the first and third cycles are subjected to acceleration and deceleration, respectively. The rates of change of displacement (both translational and angular) of the trunk body, during motion on an uneven terrain, are as shown in Fig. 3.11. The variation of body height during motion is taken care of by the translational displacement rate input along Z-axis (refer to Fig. 3.11c). Also, the roll and pitch motions are depicted by the angular displacement rate input along X and Y, respectively, (refer to Fig. 3.11d, e). The yaw motion along Z-axis is neglected, since the robot moves in a straight-forward manner.

According to the motion planning algorithms (refer to Sects. 3.2.5, 3.2.6, and 3.2.7), the position, velocity, and acceleration of the leg tip $\mathbf{P}_{i3}$ ($i = 1$ to 6) are computed with respect to frame $\boldsymbol{G}_0$. The coordinates of the leg tip $\mathbf{P}_{i3}$ are plotted in 3D Cartesian space for all the legs, as shown in Fig. 3.12. The projections show that for different cycles the support points of the legs on the terrain are at different heights which confirm that the system is moving on an uneven terrain. Also, the effect of slip has been neglected in the present study.

Inverse kinematic analysis of the robot shows the computed graphs of angular displacement, velocity, and acceleration of the joints in leg 2 for illustration (refer to Fig. 3.13). Figure 3.13a, d, g shows that the angular displacements of the joints are within the expected limits. This concludes that interference between legs could be avoided during the motion of the robot. The magnitude of angular velocities of joint 22 and 23 (refer to Fig. 3.13e, h) is higher compare to that of joint 21. Figure 3.13b

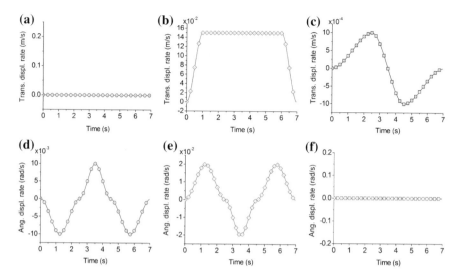

Fig. 3.11 Rate of change of displacement of a trunk body on an uneven terrain for three duty cycles with respect to frame *G* **a** translational along *X*-axis, **b** translational along *Y*-axis, **c** translational along *Z*-axis, **d** angular along *X*-axis, **e** angular along *Y*-axis, **f** angular along *Z*-axis

shows that the angular velocity of joint 21 of leg 2 during support phase is less compare to that during swing phase, which is required for close coordination of the joint motion during a cycle. Also, the angular velocities of all the joints of leg 2 varied steadily at a faster rate during swing phase compare to that during support phase. Further, during swing phase, the acceleration rates are seen to be higher and change abruptly (refer to Fig. 3.13c, f, i) compare to that during swing phase, where the variation is not much and magnitudes are low. Further, the position of the aggregate center of mass of the system during straight-forward motion with DF = 1/2 in 3D Cartesian space obtained analytically is as shown in Fig. 3.14. Though the variation in the position of COM of the robot with respect to time along *X*- and *Y*-directions (computed with respect to G_0) are small, it is not smooth, since the robot has to traverse through an uneven terrain with small undulations.

3.3.2 Case Study 2: Crab Motion of the Robot on a Banked Terrain (DF = 3/4)

Another case study is about the typical application of crab motion during maneuverability of the robot on a banked surface with DF = 3/4. The robot moves transversely on a banking surface (banking angle = 30°) and a positive crab angle (θ_c) of 70°. Unlike straight-forward motion, the *X* and *Y* (horizontal and vertical)-translational

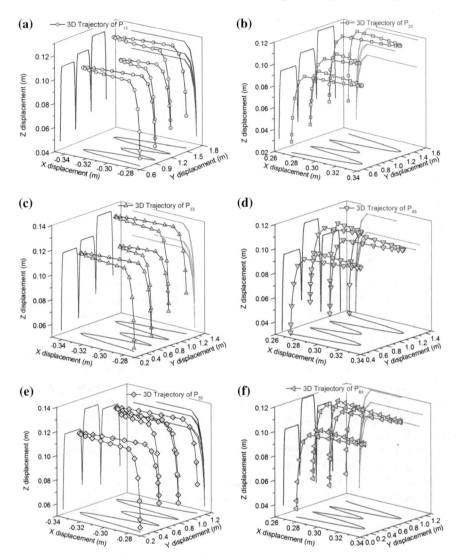

Fig. 3.12 3D motion trajectory of leg tip P_{i3} on an uneven terrain with respect to frame G_0 **a** leg 1, **b** leg 2, **c** leg 3, **d** leg 4, **e** leg 5, and **f** leg 6

displacement rates of the trunk body during crab motion are related. Hence, the translational velocity components of the trunk body with respect to frame G is assumed to follow the relationship given by Eqs. (3.64) and (3.66).

The relevant kinematic motion inputs are mentioned below: At time $t = 0$, the initial position and orientation of $\mathbf{P}_0$ with respect to frame G are given by $\mathbf{p}_0^G = \{0, 0.45, 0.15, 0, 0, 0\}^T$. Also it is assumed that when the motion of the robot starts, the value of angular velocity

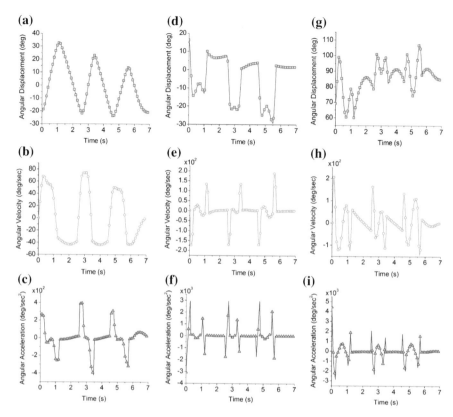

Fig. 3.13 Graphs of the kinematic analysis of a realistic six-legged robot during straight-forward motion using wave gait (DF = 1/2) on an uneven terrain for leg 2. Joint 21 **a** angular displacement, **b** angular velocity, **c** angular acceleration. Joint 22 **d** angular displacement, **e** angular velocity, **f** angular acceleration. Joint 23 **g** angular displacement, **h** angular velocity, **i** angular acceleration

and accelerations is zero, although the maximum translational and angular displacement rates are given by $\dot{\mathbf{r}}_{P_0O}^G\big|_{max} = \left(\dot{x}_{P_0O}^G, \dot{x}_{P_0O}^G \cdot \cot\theta_c, 0.001\right)^T$ where $\dot{x}_{P_0O}^G = 0.02$ m/s and $\dot{\boldsymbol{\eta}}_0\big|_{max} = \{0.01, 0.02, 0\}^T$. Further, the initial values of the joint angles of the robot follow a notation $(\theta_{i1}, \beta_{i2}, \beta_{i3})$ where $i = 1$ to 6 is ($\pm 20°, \pm 16°, \pm 69°$), respectively. The orientation of the terrain is given by $\boldsymbol{\eta}_G = (0, 30, 0)^T$. Also, the other significant inputs are associated to the topography of the terrain and measured with respect to frame L_{i3}. The values of Hm_{in} are in the order of $[Hm_{i1}, Hm_{i2}, Hm_{i3}]$ for $i = 1$ to 6 and three duty cycles ($n = 3$) such that, $[0.015, 0.015, 0.015]$ corresponds to the respective maximum heights of terrain along $\mathbf{Z}$ (Hm_{in}) the legs has to overcome during swing. Since the terrain surface is flat, $h'_{in} = 0$ for all duty cycles. The value of Δh (=0.002 m) is constant throughout. Additionally, the maximum value of slip velocity is assumed to be zero with a slip angle of $\varepsilon_s = 0°$. The lateral offset (l_i) is equal to 0.2185 m and the crab angle (θ_c) as mentioned above

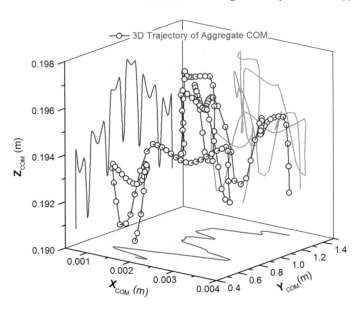

Fig. 3.14 Variation of aggregate COM of the system during locomotion on an uneven terrain with straight-forward motion in frame G_0

is equal to $70°$. The units of the different kinematic parameters, if not mentioned are as given in Sect. 3.3.

For $n = 3$, i.e., for three duty cycles simulations are tried out, with DF = 3/4, body stroke, $s_0 = 0.027$ m, and average time step (h) = 0.01 s in MATLAB. The elapsed time for three duty cycles is computed as 5.9 s with the time period of the first, second and third cycles observed as 2.05 s, 1.80 s, and 2.05 s, respectively, as computed using Eq. (3.92). It has been observed that the time for the second cycle is comparatively lesser than the time for first and last cycles. This is due to the smooth acceleration and deceleration of the trunk body during the first and last cycles, respectively, thereby, making the motion more realistic. The rates of angular and translational displacement for the trunk body during the robot's motion along the banked surface are calculated using equations given in Sect. 3.2.5.1 (refer to Fig. 3.15).

In the present case, the translational and angular displacement rates along Z-axis are assumed zero, i.e., the trunk body is always at a constant height with respect to the frame G in contrast to the previous case study in which the trunk body executes motion in an uneven terrain with varying body height. The trajectory of the tip point in 3D Cartesian space with respect to G_0 and reference axis XYZ is plotted for all the legs (refer to Fig. 3.16). Computation of the kinematic parameters (position, velocity, acceleration) is based on the motion and gait planning algorithms as discussed in Sects. 3.2.5, 3.2.6 and 3.2.7. Since the banking surface is assumed flat, no such

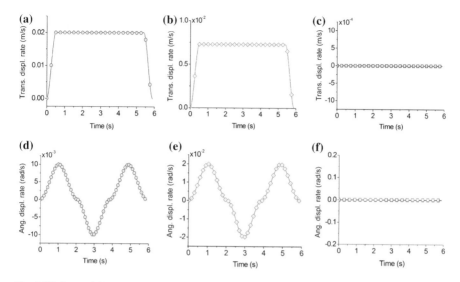

Fig. 3.15 Rate of change of displacement of a trunk body on a banking surface for three duty cycles with respect to frame **G a** translational along X-axis, **b** translational along Y-axis, **c** translational along Z-axis, **d** angular along X-axis, **e** angular along Y-axis, **f** angular along Z-axis

abrupt changes in foothold positions during the robot's motion has been observed in the projections along respective axes shown in Fig. 3.16.

Figure 3.17 shows the angular parameters (displacement, velocities and accelerations) for various joints of leg 3. It is interesting to note that angular displacement of joint 11 is not kept constant (as expected for a constant crab angle $\theta_c = 70°$) due to the presence of angular motions in the trunk body along the horizontal and vertical axes with respect to frame L_0 (refer to Fig. 3.17a). The angular displacement rates of joint 32 and 33 varied, respectively, in the range of $(-22.5°, -7.5°)$ and $(-87.5°, -45°)$ approximately (refer to Fig. 3.17d, g) and are within the expected limits. However, in crab motion interference between legs is not possible.

Further, it has been observed that due to small angular displacement variation there is a small change in the rate of angular velocity of joint 31 as shown in Fig. 3.17b. The angular velocities of the joint 32 and 33 of leg 3 varied steadily at a faster rate during swing phase compare to that during support phase which can be well interpreted from Fig. 3.17e, h. Further, during swing phase the acceleration rates are higher and changes abruptly (refer to Fig. 3.17c, f, i) in comparison with the rates during swing phase when variation is not much and magnitudes are low. Further, the position of the aggregate center of mass of the system in 3D Cartesian space during crab motion and DF $= 3/4$ is traced along XYZ-axis system with respect to G_0 (refer to Fig. 3.18). The COM variation is steady in XYZ-axis system, measured with respect to G_0.

Turning motion capabilities of the robot with the developed kinematic model has also been tested with the help of case studies described later, in Chap. 5, Sect. 5.2. The simulated results in that case have been validated with the results obtained using VP tools.

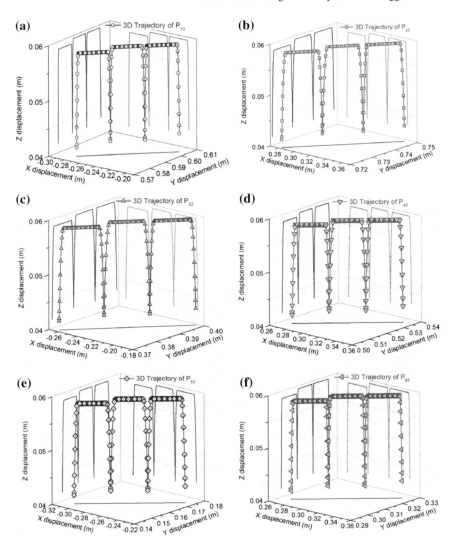

Fig. 3.16 3D motion trajectory of leg tip P_{i3} ($i = 1$ to 6) on an uneven terrain with respect to frame G_0 **a** leg 1, **b** leg 2, **c** leg 3, **d** leg 4, **e** leg 5, and **f** leg 6

3.4 Summary

In this work, a classical approach has been applied to develop a generalized kinematic model of six-legged robot and simulation of the motions and mechanisms in 3D Cartesian space has been developed. The kinematic model is capable enough of carrying out complex motion analysis (such as straight-forward, crab, turning, etc.) of the robotic system negotiating various terrains (even to uneven, flat level to

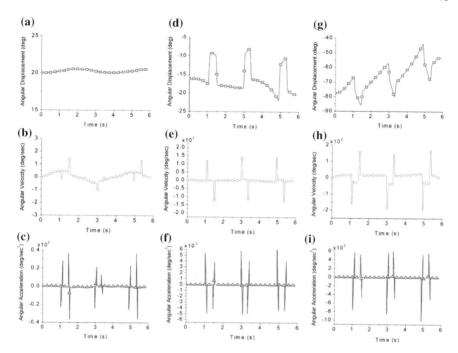

Fig. 3.17 Graphs of the kinematic analysis of a realistic six-legged robot during crab motion on a banking surface using wave gait (DF = 3/4) for leg 3. Joint 31 **a** angular displacement, **b** angular velocity, **c** angular acceleration. Joint 32 **d** angular displacement, **e** angular velocity, **f** angular acceleration. Joint 33 **g** angular displacement, **h** angular velocity, **i** angular acceleration

elevated, and others) with the parameters like support foot-tip slippage condition; varying trunk body height; trunk body velocity with roll, pitch, and yaw motions; body stroke; lateral offset; crab angle, etc. The performance of the system has been evaluated through computer simulations, thereby, obtaining all the feasible solutions corresponding to displacement, velocity, acceleration, aggregate COM of the system at any instant of time. The developed algorithms for motion planning (of both trunk body and swing leg trajectory planning) have substantial impact on the inverse kinematic analysis of the closed chain formed during the stance phase of the feet relative to the trunk body motion or open chains formed during the swing phase of the feet in such a spatially complex environment. Furthermore, the developed constraint equations can be used as the direct aid to study the constrained, coupled multi-body dynamics of the robot with impact and slip phenomena.

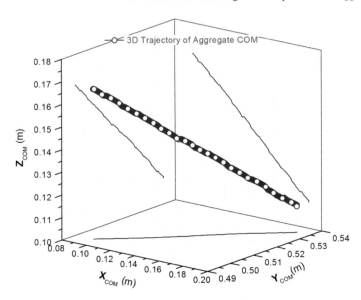

Fig. 3.18 Variation of Aggregate COM of the system during locomotion on a banked surface (banking angle, $\beta_G = 30°$) with crab motion in frame G_0

References

1. H. Hahn, *Rigid Body Dynamics of Mechanisms*, 1st edn. (Springer, Berlin, Heidelberg, 2002)
2. A. Mahapatra, S.S. Roy, D.K. Pratihar, Computer aided modeling and analysis of turning motion of hexapod robot on varying terrains. Int. J. Mech. Mater. Des. **11**(3), 309–336 (2015)
3. MSC.ADAMS Documentation and help user guide (2010). https://simcompanion.mscsoftware. com/infocenter. MSC Software Corporation
4. P.E. Nikravesh, *Computer-Aided Analysis of Mechanical Systems* (Prentice Hall, Englewood Cliffs, NJ, 1988)
5. S.M. Song, K.J. Waldron, *Machines That Walk: The Adaptive Suspension Vehicle* (The MIT Press, Cambridge, 1989)
6. S. Tickoo, D. Maini, V. Raina, *CATIA V5R16 for Engineers and Designers* (Dreamtech Press, New Delhi, 2007)

Chapter 4
Multi-body Inverse Dynamic Modeling and Analysis of Six-Legged Robots

In this chapter, investigations conforming to the optimal feet forces' distributions under body force and foot–ground interactions for a realistic robot locomoting on various terrains have been explained. The constrained inverse dynamics model is formulated as a coupled dynamical problem using Newton–Euler (NE) approach in Cartesian coordinates. Kinematic transformations approach is implemented, whereby the robot kinematics in Cartesian space with large implicit constraints is transformed to joint space with a reduced explicit set of constrained equations to tackle the complexities and make the computation less intensive. The chapter also proposes a three-dimensional (3D) foot–ground interaction mechanics model with the compliance contact. The model is categorized as the deformable foot on hard terrain (little deformation) and developed based on (a) compliant contact force model to describe foot–ground collision and (b) Amonton–Coulomb's friction model to describe stick and slip phenomenon. Further, energy consumption and stability measures are also carried out in the present work. The efficacy of the developed dynamic model was tested using few case studies.

4.1 Analytical Framework

The present section deals with the analytical modeling of realistic hexapedal robot, and hence, is applicable to any other legged robot design. It forms both closed and open chain kinematic loops creating a complex mechanical system. When the legs are coupled to one another through a trunk body and also through the ground, it forms a closed chain kinematics loop and the legs are said to be in the support phase. Likewise, for an open chain kinematics loop, one of the legs is in swing phase and is connected to one another only through the trunk body. Forces and moments transmit through these kinematic chains from one leg to another, and hence, dynamic coupling exists. The dynamics equations of motion of such a complex mechanism of the hexapedal system maneuvering in various terrains are described in the following sections by taking into consideration multiple reference systems attached to it in

© Springer Nature Singapore Pte Ltd. 2020
A. Mahapatra et al., *Multi-body Dynamic Modeling of Multi-legged Robots*,
Cognitive Intelligence and Robotics, https://doi.org/10.1007/978-981-15-2953-5_4

Cartesian coordinates (both global and local) as discussed in Sect. 3.2.1. Some of the important assumptions made in the book are as follows:

1. All the kinematic link lengths are rigid.
2. The contact area of leg tip and subsequent deformation is negligibly small. Also, inertial forces due to the deformation are negligible.
3. The Coulomb's friction model is assumed to follow cubic polynomials for smooth transitions. The model is based on stiction to transition velocity and *vice versa* and is related to static and dynamic coefficients of friction, respectively.
4. Joint frictions are neglected.
5. Rebound on impact is neglected.

4.1.1   Implicit Constrained Inverse Dynamic Model

Multi-body dynamical models for legged robot systems are typically characterized by many DOF, contact constraints or collision events, variety of potential feet–ground interaction models and mass-inertial parameter settings due to changing load conditions. Also, for a multi-body system like a hexapedal robot, complexities in the dynamical model arise due to the dynamic coupling caused by the kinematic constraints in each loop (formed by the support legs and the trunk body) and overdeterminate input mechanisms. Such dynamic models are required for the realistic simulation of legged system behaviors and, thereby, fine-tuning the gait parameters, geometrical design parameters, and feedback control, etc., for real-life robots. The present problem uses recursive Newton–Euler approach to carry out the inverse dynamic or kinetostatic analysis of the system (Bastos et al. [2]; Bessonnet et al. [4]; Otten [29]. This means that a particular motion of the system is sought, and the purpose is to find out the forces and moments that must work on the system to generate such a motion.

The dynamic modeling of a spatial hexapedal robotic structure is a vast problem that leads to a huge number of equations. In the present case, there are 114 ($= 6n_b$, where $n_b = 19$, is the number of rigid bodies of the system) implicit constrained dynamic equations in the system. They are expressed in over complete Cartesian coordinates $(\mathbf{p}^{G_0})$ in frame $\mathbf{G}_0$, such that

$$\mathbf{M}(\mathbf{p}^{G_0}) \cdot \dot{\mathbf{v}}^{G_0} = {}^{cf} + \mathbf{f}(\mathbf{p}^{G_0}, \mathbf{v}^{G_0}) + \mathbf{q}_{GC}(\mathbf{p}^{G_0}, \mathbf{v}^{G_0}) \in \mathbb{R}^{114} \qquad (4.1)$$

$$ {}^{cf} = (\mathbf{g}_{\mathbf{p}}(\mathbf{p}^{G_0}) \cdot \mathbf{T}(\mathbf{p}^{G_0}))^{\mathbf{T}} \cdot \boldsymbol{\lambda}; \in \mathbb{R}^{114} \qquad (4.2)$$

$$\mathbf{f}(\mathbf{p}^{G_0}, \mathbf{v}^{G_0}) = \mathbf{f}_K + \mathbf{f}_U; \in \mathbb{R}^{114} \qquad (4.3)$$

Table 4.1 Mass moment of the inertia of the six-legged robot

Components	Mass moment of inertia ($\times 10^{-6}$ kg m^2)					
	J_{xx}	J_{yy}	J_{zz}	J_{xy}	J_{yz}	J_{zx}
Trunk body	16652.9	2518.5	16897.2	0	0	0
Payload	8524.2	3430.9	10823.2	0	0	0
Link $i1$	70.83	108.38	56.74	-3.50	-2.18	-4.96
Link $i2$	20.19	86.70	100.26	12.9	1.0	-7.1
Link $i3$	98.45	87.52	20.78	1.23×10^{-4}	1.73	-1.95×10^{-5}

NB $J_{xx} \equiv J_{C_B x}^{L_0}$ or $J_{C_L x}^{L_0}$ or $J_{C_{ij} x}^{L'_{ij}}$; $J_{yy} \equiv J_{C_B y}^{L_0}$ or $J_{C_L y}^{L_0}$ or $J_{C_{ij} y}^{L'_{ij}}$ and so on. Refer to Appendix A.8

where

$$\mathbf{v}^{G_0} = \left((\mathbf{v}_0^{G_0})^{\mathrm{T}}, (\mathbf{v}_1^{G_0})^{\mathrm{T}}, (\mathbf{v}_2^{G_0})^{\mathrm{T}}, (\mathbf{v}_3^{G_0})^{\mathrm{T}}, (\mathbf{v}_4^{G_0})^{\mathrm{T}}, (\mathbf{v}_5^{G_0})^{\mathrm{T}}, (\mathbf{v}_6^{G_0})^{\mathrm{T}} \right)^{\mathrm{T}} \in \mathbb{R}^{114} \quad (4.4)$$

and

$$\dot{\mathbf{v}}^{G_0} = \left((\dot{\mathbf{v}}_0^{G_0})^{\mathrm{T}}, (\dot{\mathbf{v}}_1^{G_0})^{\mathrm{T}}, (\dot{\mathbf{v}}_2^{G_0})^{\mathrm{T}}, (\dot{\mathbf{v}}_3^{G_0})^{\mathrm{T}}, (\dot{\mathbf{v}}_4^{G_0})^{\mathrm{T}}, (\dot{\mathbf{v}}_5^{G_0})^{\mathrm{T}}, (\dot{\mathbf{v}}_6^{G_0})^{\mathrm{T}} \right)^{\mathrm{T}} \in \mathbb{R}^{114} \quad (4.5)$$

are the velocity and acceleration vectors of the system, respectively; $^c\mathbf{f} \in \mathbb{R}^{114}$ being the vector of constraint reaction forces and torques of the joints linked with system coordinates; $\mathbf{f}(\mathbf{p}^{G_0}, \mathbf{v}^{G_0})$ is the vector of both known and unknown applied forces and torques denoted by $\mathbf{f}_K$ and $\mathbf{f}_U$, respectively; $\mathbf{q}_{GC}(\mathbf{p}^{G_0}, \mathbf{v}^{G_0}) \in \mathbb{R}^{114}$ is the vector of centrifugal forces and gyroscopic terms; $\mathbf{g}_\mathbf{p}(\mathbf{p}^{G_0}) \cdot \mathbf{T}(\mathbf{p}^{G_0}) \in \mathbb{R}^{n_c, 114}$ is the constraint Jacobian matrix; $\boldsymbol{\lambda} \in \mathbb{R}^{n_c}$ is the vector of the Lagrange multipliers; $\mathbf{M}(\mathbf{p}^{G_0}) \in \mathbb{R}^{114, 114}$ is the combined mass matrix of the robotic system. The mass moment of inertias of the main components of the six-legged robot are given in Table 4.1.

The set of nonlinear Eqs. (3.16), (4.1), (4.2), and (4.3) is called the system of differential algebraic equations (DAEs). These DAEs govern the state of the robot with respect to frame G_0 at any instant of time.

In rigid multi-body systems like six-legged robots, it is often difficult to model the complex structure. Hence, one has to opt for methods like the free-body diagram (FBD) to model the dynamics of system [1]. In the present case, following this approach, the Newton–Euler equations are obtained for every rigid body forming the total system as shown in Fig. 3.1a. The total system in the present context means the robotic structure formed by the nineteen rigid bodies: (a) trunk body, (b) coxa (link $i1$, $i = 1$–6), (c) femur (link $i2$, $i = 1$–6), and (d) tibia (link $i3$, $i = 1$–6). Each of the bodies is isolated to build the FBD diagram with all the external forces and torques/moments, as shown in Fig. 4.1. The external forces to the structure are the gravity and friction forces on the feet tips (joint frictions are neglected). Further, the external torques/moments are the actuator torques/moments on feet tips. The forces

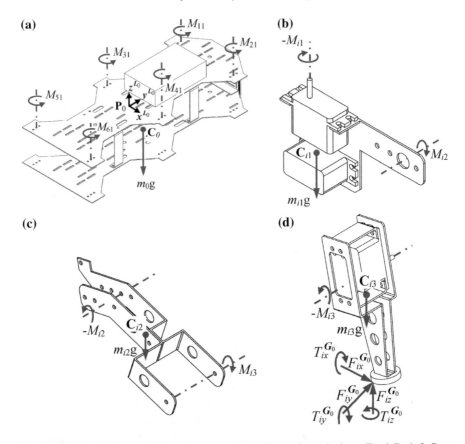

Fig. 4.1 Free-body diagram of the main components of the six-legged robot **a** Trunk Body **b** Coxa (link $i1$), **c** Femur (link $i2$), **d** Tibia (link $i3$) (*workless constraint reaction forces, inertia forces, and torques are not shown in figure*)

and torques/moments that are internal to the system are handled in an explicit way, along with the inertial and Coriolis forces. The Newton–Euler approach applies the vectorial dynamic equations to each of the isolated bodies of the system discussed in Appendix A.8. The overall set is achieved by joining all the element equations.

4.1.2 Newtonian Mechanics with Explicit Constraints

The implicit constrained dynamics Eq. (4.1) expressed in overcomplete Cartesian coordinates $(\mathbf{p}^{G_0})$ is often undesirable due to a large number of equations, which makes the problem computationally intensive. Hence, it is better to express the dynamics in generalized coordinates ($\mathbf{q}$). A kinematic transformation is carried out to describe the position and velocity components in terms of $\mathbf{q}$ Bennani and Giri [3].

This is done through explicit constraints given by the forward kinematic function,

$$\mathbf{p}^{G_0} = \mathbf{h}(\mathbf{q}) \in \mathbb{R}^{114} \tag{4.6}$$

where

$$\mathbf{q} = \left((\mathbf{q}_0)^{\mathrm{T}}, (\mathbf{q}_1)^{\mathrm{T}}, (\mathbf{q}_2)^{\mathrm{T}}, (\mathbf{q}_3)^{\mathrm{T}}, (\mathbf{q}_4)^{\mathrm{T}}, (\mathbf{q}_5)^{\mathrm{T}}, (\mathbf{q}_6)^{\mathrm{T}} \right)^{\mathrm{T}} \in \mathbb{R}^{24} \tag{4.7}$$

$$\mathbf{q}_0 \equiv \mathbf{p}_0^{G_0} = \left((\mathbf{r}_{P_0 O}^{G_0})^{\mathrm{T}}, (\mathbf{\eta}_0)^{\mathrm{T}} \right)^{\mathrm{T}} \in \mathbb{R}^6 \tag{4.8}$$

$$\mathbf{q}_i = (\theta_{i1}, \beta_{i2} \beta_{i3})^{\mathrm{T}} \quad i = 1\text{--}6 \in \mathbb{R}^{18} \tag{4.9}$$

Here, the twenty-four generalized coordinate vectors $\mathbf{q} \in \mathbb{R}^{24}$ are chosen as six spatial coordinates of the trunk body and eighteen rotational coordinates of the joint angles of the system. Refer to Appendix A.9.

The vector of velocity constraint equations of the system that relates the Cartesian space velocities $\left(\mathbf{v}^{G_0} \in \mathbb{R}^{24} \right)$ and joint space velocities $\left(\mathbf{u} \in \mathbb{R}^{24} \right)$ is given by the expression:

$$\mathbf{v}^{G_0} = \mathbf{J} \cdot \mathbf{u} \in \mathbb{R}^{114} \tag{4.10}$$

Differentiating (4.10) with respect to time gives the expression of the acceleration constraint equations of the system such that

$$\dot{\mathbf{v}}^{G_0} = \mathbf{J} \cdot \dot{\mathbf{u}} + \dot{\mathbf{J}} \cdot \mathbf{u} \in \mathbb{R}^{114} \tag{4.11}$$

where $\mathbf{J} \in \mathbb{R}^{114,24}$ is the Jacobian matrix of the system in terms of $\mathbf{q}$, linked to ground reaction forces and coupled joint torques as discussed later. Also, it is to be noted that multiplication of the term ${}^c\mathbf{f}$ by Jacobian matrix $\mathbf{J}$ eliminates the constraints forces,

$$\mathbf{J}^T \cdot {}^c\mathbf{f} = \mathbf{0} \tag{4.12}$$

Therefore, the transformation of the inverse dynamic model (in terms of $\mathbf{q}$) is realized by considering the following relation:

$$\mathbf{D}(\mathbf{q})\dot{\mathbf{u}} + \mathbf{C}(\mathbf{q}, \dot{\mathbf{q}}) = \mathbf{\tau}, \in \mathbb{R}^{24} \tag{4.13}$$

where $\mathbf{D}(\mathbf{q})$ is the coupled mass and inertia matrix of the robotic system expressed in terms of generalized coordinates, $\mathbf{\tau}$ is the vector of ground reaction forces/moments and coupled joint torques.

Also,

$$\mathbf{D}(\mathbf{q}) = \mathbf{J}^{\mathrm{T}} \mathbf{M}(\mathbf{p}^{G_0}) \mathbf{J} \in \mathbb{R}^{24 \times 24} \tag{4.14}$$

$$\mathbf{C}(\mathbf{q}, \dot{\mathbf{q}}) = \mathbf{J}^{\mathrm{T}}\mathbf{M}(\mathbf{p}_0^G)\dot{\mathbf{J}}\mathbf{u} \in \mathbb{R}^{24} \tag{4.15}$$

$$\boldsymbol{\tau} = \mathbf{J}^{\mathrm{T}}\left[\mathbf{f}(\mathbf{p}^{G_0}, \mathbf{v}^{G_0}) + \mathbf{q}_{GC}(\mathbf{p}^{G_0}, \mathbf{v}^{G_0})\right] \in \mathbb{R}^{24} \tag{4.16}$$

Twenty-four numbers of equations are obtained in terms of ground reaction forces/moments and coupled joint torques by substituting the necessary input values in Eq. (4.13). The equations are further re-arranged as described in the following paragraph.

The first set of six equations of (4.13) describes the dynamic behavior of the trunk body and payload (combined) at any instant of time with respect to forces and moments such that

$$\sum \mathbf{F}_i^{G_0} + \mathbf{F}_e^{G_0} = \mathbf{0}_3 \in \mathbb{R}^3 \tag{4.17}$$

$$\sum (\mathbf{s}_i^{G_0} \times \mathbf{F}_i^{G_0}) + \mathbf{M}_0^{G_0} + \mathbf{M}_e^{G_0} = \mathbf{0}_3 \in \mathbb{R}^3 \tag{4.18}$$

The following set of eighteen equations of (4.13) describes the correlations between the joint torques and ground reaction forces and moments on the legs, so that for leg i,

$$\mathbf{M}_i^{G_0} = -\mathbf{B}_i^{-1}\left(\mathbf{A}_i\mathbf{F}_i^{G_0} + \mathbf{D}_i\mathbf{T}_i^{G_0} + \mathbf{M}_{ei}^{G_0}\right) \in \mathbb{R}^3 \tag{4.19}$$

where $\mathbf{M}_i^{G_0}$ is the vector of joint torques of leg i denoted by $[M_{i1}^{G_0}\ M_{i2}^{G_0}\ M_{i3}^{G_0}]^{\mathrm{T}}$; $\mathbf{M}_{ei}^{G_0}$ is the vector representing the centrifugal, Coriolis, gyroscopic, and gravitational moments acting on the leg i denoted by $[M_{ei1}^{G_0}\ M_{ei2}^{G_0}\ M_{ei3}^{G_0}]^{\mathrm{T}}$; $\mathbf{B}_i = [1\ 0\ 0; 0\ 1\ 1; 0\ 0\ 1]^{\mathrm{T}}$; $\mathbf{A}_i$ and $\mathbf{D}_i$ are the square matrices denoted by $[a_{i1x}\ a_{i1y}\ a_{i1z}; a_{i2x}\ a_{i2y}\ a_{i2z}; a_{i3x}\ a_{i3y}\ a_{i3z}]^T$ and $[d_{i1x}\ d_{i1y}\ d_{i1z}; d_{i2x}\ d_{i2y}\ d_{i2z}; d_{i3x}\ d_{i3y}\ d_{i3z}]^T$, respectively; $\mathbf{s}_i^{G_0}$ is the displacement vector from point $\mathbf{P}_0$–$\mathbf{P}_{i3}$ represented in reference frame G_0; $\mathbf{M}_0^{G_0}$ is the vector of joint torques acting on the trunk body and payload (combined) denoted by $\left[0\ 0\ \sum_{i=1}^{6} M_{i1}^{G_0}\right]^{\mathrm{T}}$; $\mathbf{F}_e^{G_0}$ and $\mathbf{M}_e^{G_0}$ are the vectors representing centrifugal, Coriolis, gyroscopic, and gravitational forces and moments, respectively, acting on the combined mass of the trunk body and payload; $\mathbf{F}_i^{G_0}$ and $\mathbf{T}_i^{G_0}$ are the ground reaction forces and moments at the foot of support leg i with respect to G_0 denoted by vector $[F_{ix}^{G_0}\ F_{iy}^{G_0}\ F_{iz}^{G_0}]^{\mathrm{T}}$ and $[T_{ix}^{G_0}\ T_{iy}^{G_0}\ T_{iz}^{G_0}]^{\mathrm{T}}$, respectively, discussed in Sect. 3.1.3.2. Further, for swing legs, $\mathbf{F}_i^{G_0} = \mathbf{0}_3$; $\mathbf{T}_i^{G_0} = \mathbf{0}_3$.

4.1.3 Three-Dimensional Contact Force Model

A legged robot makes contact with its environment during execution of its tasks. Therefore, contact modeling is an unavoidable part of the studies in this field and is classified into two types: rigid and compliant. Rigid body contact models are developed mainly based on an assumption of 'point contact,' although it is impossible for real bodies to sustain nonzero impact forces at one point. As contact occurs over a certain region in 3D space, rigid body contact models are not accurate to model such impacts. The most appropriate models for such interaction mechanics are known as compliant (viscoelastic) contact force models. Using these models, various types of terrains can be simulated by lumped parameters with appropriate characteristics [37, 22]. Moreover, such contact models are capable of taking into account the nonlinear characteristics involved in the interface, thereby mimicking some level of natural dynamics and/or assisting in force control. In the present book, a foot–ground interaction model has been proposed. It is based on deformable foot on hard terrain with little deformation. The model defines the nonlinear interactive forces and moments with respect to viscoelastic compliant contact force theory and Amonton–Coulomb's friction law.

4.1.3.1 Compliant Contact Impact Model

It is to be noted that the foot characteristic of the robot's leg (that includes geometrical shape, dimensions, material, etc.) has significant effect on the mechanics of interaction. In the current model, the foot of the robot is assumed to be made of rubber material which has embedded spring and damping characteristics. These characteristics influence the interaction mechanics of the foot with the terrain when it makes an oblique impact. Further, the following assumptions are made:

(a) Distributed contact force on the impacting bodies is substituted by an effective force at a point. This point is designated as $\mathbf{P}'_{i3}$ (initial point of contact). It is to be remembered that this distributive contact force acts in an infinitesimally small contact area which is much less than the characteristic length of the colliding pad (refer to Fig. 4.2).

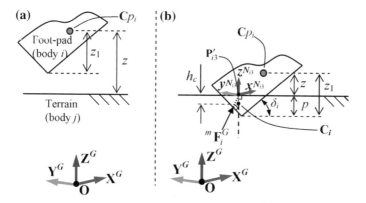

Fig. 4.2 Foot–terrain interaction model **a** before impact **b** after impact (Cp_i-COM of the footpad)

(b) Both the foot and terrain are considered as rigid except the interacting zone surrounding the contact point. The interacting zone has a small contact volume with COM at C_i where compliancy between the contacting bodies exists.

(c) Contact zone is administered by a viscoelastic model, and energy losses due to elastic vibrations are small.

Figure 4.2 shows the foot–terrain interaction process that occurs between the footpad and the terrain. It is to be noted that when z is less than z_1, contact detection is initiated. As soon as contact is detected, the centroid (C_i) of the engulfed volume is calculated. Subsequently, the points (P'_{i3} and D) which are the closest points, respectively, from each of the body's surface from C_i are computed (refer to Fig. 4.3a, b). Here, the points P'_{i3} and D are the meeting points of the perpendicular lines $C_iP'_{i3}$ and AB as well as C_iD and HB, respectively. The line joining the two P'_{i3} and D is the line of oblique impact force between the feet tip and terrain. Further, line $P'_{i3}D$ is the line along which the resultant impact force ${}^m F_i^G$ acts as shown in Fig. 4.3b. From Fig. 4.3a, it is also seen that the Z-axis of the local frame N_{i3} is perpendicular to the terrain surface, whereas the Y-axis is aligned with the longitudinal direction

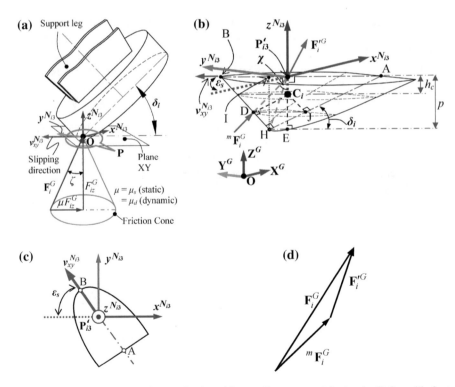

Fig. 4.3 Foot–terrain interaction mechanics with compliance contact for leg i **a** Deformable foot on hard terrain (with negligible deformation) during support phase, **b** Close view 'P' of the 3D inter-action area of the foot tip with resultant compliant contact and ground reaction forces, **c** Direction of foot slip (top view), **d** Force vector diagram

of the motion of the robot. The point C_i lies on the plane ABH and is at a depth, h_c (COM of the contact volume) from the plane XY. χ is the angle subtended by the impact force $^m\mathbf{F}_i^G$ with the point C_i, that is, $\angle C_i P'_{i3} D$. Further, the lines **AB**, **DJ**, and **HE** are parallel and lie on a common plane. The point **I** lies on the line joining **AB**, which is the direction of leg slip with origin at point $\mathbf{P}'_{i3}$. Refer to Appendix A.10 for trigonometrical relationships.

When the feet tip impacts obliquely on the terrain, the resultant compliant impact force $^m\mathbf{F}_i^G$ (refer to Fig. 4.2b and Fig. 4.3b) with respect to frame G is resolved into three components (along X-, Y-, and Z-axis, respectively) acting at $\mathbf{P}'_{i3}$ and is expressed as follows:

$$^m\mathbf{F}_i^G = \left[\, ^m F_{ix}^G \; ^m F_{iy}^G \; ^m F_{iz}^G \, \right]^{\mathrm{T}}. \tag{4.20}$$

The compliant normal impact force $^m F_{iz}^G$ for a deformable foot (hard) and hard (little deformation) terrain follows an impact function. The function consists of elastic force and damping force. Therefore, $^m F_{iz}^G$ depends mainly on the foot's deformation, its velocity, contact stiffness, damping, etc. (Liang and Xin [19] The impact-based contact model in the present book is based on the nonlinear Hunt–Crossley model (Hunt and Crossley [15] with the introduction of some modifications (MSC.ADAMS Documentation [28]). Hence, the compliant normal impact force is given by

$$^m F_{iz}^G = \begin{cases} 0 & z > z_1 \\ K(z_1 - z)^e - C_{\max}.\dot{z}.f(z) & z \le z_1 \end{cases} \tag{4.21}$$

where $f(z)$ is a MSC.ADAMS$^®$ step function that follows a cubic polynomial such that

$$f(z) = \mathrm{STEP}(z, z_1 - p, 1, z_1, 0) \equiv 1 - \bar{a}_z.\Delta_z^2(3 - 2\Delta_z) \tag{4.22}$$

$$\bar{a}_z = 1; \tag{4.23}$$

$$\Delta_z = (z - z_1 + p)\big/ p; \tag{4.24}$$

Here, z is the distance function, z_1 is the trigger distance, and $\dot{z}$ is the derivative of z to impact. It is also be seen that when the distance between the two objects is less than the free length of z (refer to Fig. 4.2) that is when $z \le z_1$, the impact function activates. Also, e is the contacting force index (generally material property) where $e > 1$ for stiff spring. Again, K denotes the characteristic stiffness of contact that depends on material and shape of the impacting bodies considered in MSC.ADAMS$^®$, $C_{\max}$ represents the maximum damping, which is 1% of K (*thumb rule*), p is the damping ramp up distance ($\ll z_1$), that is, boundary penetration at which MSC.ADAMS$^®$ solver applies full damping (refer to Fig. 4.2). However, in practice it is very difficult to tune the stiffness (K) and damping parameters (C_{max}) of such contact models, and hence, causes numerical difficulties for multi-contact systems.

Subsequently, the force becomes nonzero and can be divided into two parts: an exponential spring force and a damping force that follow a step function (Mahapatra et al. [24]. Both these forces are strictly positive in nature, and hence, oppose the compression that occurs during penetration.

Further, the resultant of the components of the tangential compliance contact impact forces ($^mF_{ix,}\ ^mF_{iy}$) in the region is given by $^mF_{ixy}$ such that

$$^m F^G_{ixy} = \sqrt{\left(^m F^G_{ix}\right)^2 + \left(^m F^G_{iy}\right)^2} \qquad (4.25)$$

It is to be noted that force $^mF^G_{ixy}$ acts along the line $\mathbf{P'}_{i3}\mathbf{A}$ whose direction is opposite to the direction of slip velocity $v^{N_{i3}}_{xy}$, which acts at an angle ε_s with the X-axis of the local frame N_{i3} (refer to Fig. 4.3c).

4.1.3.2   Interactive Forces and Moments

During landing of foot on the terrain, if there is zero impact, the resultant ground reaction force in the foot of leg i with respect to frame $\mathbf{G}$ is given by

$$\mathbf{F}'^G_i = \left[\ F'^G_{ix}\ \ F'^G_{iy}\ \ F'^G_{iz}\ \right]^{\mathrm{T}} \qquad (4.26)$$

When there is impact during landing as discussed in Sect. 4.1.3.1, the force $^m\mathbf{F}^G_i$ can be expressed vectorially as follows:

$$^m\vec{\mathbf{F}}^G_i = \hat{\mathbf{n}}.^m F^G_i, \qquad (4.27)$$

where $\hat{\mathbf{n}}$ is the unit vector along the direction $\mathbf{DP'}_{i3}$, (refer to Fig. 4.3) and is given by

$$\mathbf{n} = \left[\ \sin\chi\,\cos\varepsilon_s\ \ -\sin\chi\,\sin\varepsilon_s\ \ \cos\chi\ \right]^{\mathrm{T}}. \qquad (4.28)$$

Substituting Eq. (4.28) in (4.27), we get

$$^m\vec{\mathbf{F}}^G_i = {}^m F^G_i \sin\chi\,\cos\varepsilon_s\hat{\mathbf{i}} - {}^m F^G_i \sin\chi\,\sin\varepsilon_s\hat{\mathbf{j}} + {}^m F^G_i \cos\chi\hat{\mathbf{k}}. \qquad (4.29)$$

Again, from Fig. 4.3(b), it can be stated that

$$^m F^G_{ixy} = {}^m F^G_i \sin\chi, \qquad (4.30)$$

$$^m F^G_{iz} = {}^m F^G_i \cos\chi. \qquad (4.31)$$

Substituting Eqs. (4.30) and (4.31) in Eq. (4.29),

$$^{m}\vec{\mathbf{F}}_{i}^{G} = {}^{m}F_{ixy}^{G}\cos\varepsilon_{s}\hat{\mathbf{i}} - {}^{m}F_{ixy}^{G}\sin\varepsilon_{s}\hat{\mathbf{j}} + {}^{m}F_{iz}^{G}\hat{\mathbf{k}}., \tag{4.32}$$

Again from Eqs. (4.30) and (4.31),

$$^{m}F_{ixy}^{G} = {}^{m}F_{iz}^{G}\tan\chi. \tag{4.33}$$

Substituting Eq. (4.33) in Eq. (4.32), we get

$$^{m}\vec{\mathbf{F}}_{i}^{G} = {}^{m}F_{iz}^{G}\tan\chi\cos\varepsilon_{s}\hat{\mathbf{i}} - {}^{m}F_{iz}^{G}\tan\chi\sin\varepsilon_{s}\hat{\mathbf{j}} + {}^{m}F_{iz}^{G}\hat{\mathbf{k}}, \tag{4.34}$$

or

$$^{m}\mathbf{F}_{i}^{G} = \left[{}^{m}F_{iz}^{G}\tan\chi\cos\varepsilon_{s} \ -{}^{m}F_{iz}^{G}\tan\chi\sin\varepsilon_{s} \ {}^{m}F_{iz}^{G} \right]^{\mathrm{T}}. \tag{4.35}$$

By comparing (4.35) and (4.20), we get

$$^{m}F_{ix}^{G} = {}^{m}F_{iz}^{G}\tan\chi\cos\varepsilon_{s}, \tag{4.36}$$

$$^{m}F_{iy}^{G} = -{}^{m}F_{iz}^{G}\tan\chi\sin\varepsilon_{s}, \tag{4.37}$$

The resultant reaction force vector (refer to Fig. 4.3d) acting at foot tip during impact with respect to **G** is given by

$$\vec{\mathbf{F}}_{i}^{G} = \vec{\mathbf{F}}_{i}^{\prime G} + {}^{m}\vec{\mathbf{F}}_{i}^{G}. \tag{4.38}$$

Again, substituting Eqs. (4.20) and (4.26), in (4.38),

$$\mathbf{F}_{i}^{G} = ({}^{m}F_{ix}^{G} + F_{ix}^{\prime G})\cdot\hat{\mathbf{i}} + ({}^{m}F_{iy}^{G} + F_{iy}^{\prime G})\cdot\hat{\mathbf{j}} + ({}^{m}F_{iz}^{G} + F_{iz}^{\prime G})\cdot\hat{\mathbf{k}}, \tag{4.39}$$

or

$$\mathbf{F}_{i}^{G} = \left[F_{ix}^{G} \ F_{iy}^{G} \ F_{iz}^{G} \right]^{\mathrm{T}}, \tag{4.40}$$

where

$$F_{ix}^{G} = {}^{m}F_{ix}^{G} + F_{ix}^{\prime G}, \tag{4.41}$$

$$F_{iy}^{G} = {}^{m}F_{iy}^{G} + F_{iy}^{\prime G}, \tag{4.42}$$

$$F_{iz}^{G} = {}^{m}F_{iz}^{G} + F_{iz}^{\prime G}. \tag{4.43}$$

Thereafter, transforming the net ground reaction force vector $\mathbf{F}_i^G$, (i.e., with respect to frame G) at the foot of leg i with respect to frame G_0, the expression obtained is as follows:

$$\mathbf{F}_i^{G_0} = \mathbf{A}^{G_0 G}\mathbf{F}_i^G = \mathbf{A}^{G_0 G}\mathbf{F}_i'^G + \mathbf{A}^{G_0 G m}\mathbf{F}_i^G = \mathbf{F}_i'^{G_0} + {}^m\mathbf{F}_i'^{G_0}, \qquad (4.44)$$

where $\mathbf{A}^{G_0 G} \in \mathbb{R}^{3,3}$ is the transformation matrix (refer to Appendix A.3 that maps reference frame (G) into reference frame (G_0).

Further, during interaction of foot with the terrain, interactive moments are also developed. They are approximated by the products of the resultant reaction force and their related arms. Therefore, moments about a point $\mathbf{H}$ (refer to Fig. 4.3b) are given by

$$\vec{\mathbf{T}}_i = \overrightarrow{\mathbf{HP}}'_{i3} \times \vec{\mathbf{F}}_i = \vec{\mathbf{r}}_i \times \vec{\mathbf{F}}_i, \qquad (4.45)$$

Therefore, with respect to frame N_{i3}

$$\vec{\mathbf{T}}_i^{N_{i3}} = \vec{\mathbf{r}}_i^{N_{i3}} \times \vec{\mathbf{F}}_i^{N_{i3}}, \qquad (4.46)$$

or

$$\mathbf{T}_i^{N_{i3}} = \tilde{\mathbf{r}}_i^{N_{i3}} \cdot \mathbf{F}_i^{N_{i3}}. \qquad (4.47)$$

where $\vec{\mathbf{r}}_i^{N_{i3}}$ is the displacement vector represented in frame N_{i3} from point $\mathbf{P}'_{i3}$ to H. Refer to Appendix A.10 to calculate the coordinates of the vector $\vec{\mathbf{r}}_i^{N_{i3}}$.

Hence,

$$\mathbf{T}_i^G = \mathbf{A}^{GN_{i3}}\mathbf{T}_i^{N_{i3}}, \qquad (4.48)$$

and

$$\mathbf{T}_i^{G_0} = \mathbf{A}^{G_0 G}\mathbf{T}_i^G = \mathbf{A}^{G_0 G}\mathbf{A}^{GN_{i3}}\mathbf{T}_i^{N_{i3}} = \mathbf{A}^{G_0 N_{i3}}\tilde{\mathbf{r}}_i^{N_{i3}}\mathbf{F}_i^{N_{i3}}, \qquad (4.49)$$

where $\mathbf{A}^{G_0 N_{i3}} \in \mathbb{R}^{3,3}$ is the transformation matrix (refer to Appendix A.3 that maps reference frame (N_{i3}) into reference frame (G_0)). Since, it is assumed that G and N_{i3} are parallel,

$$\mathbf{F}_i^{N_{i3}} = \mathbf{F}_i^G, \qquad (4.50)$$

Substituting Eqs. (4.50) and (4.44) in Eq. (4.49) gives

$$\mathbf{T}_i^{G_0} = \mathbf{A}^{G_0 N_{i3}}\tilde{\mathbf{r}}_i^{N_{i3}}\mathbf{F}_i^G = \mathbf{A}^{G_0 N_{i3}}\tilde{\mathbf{r}}_i^{N_{i3}}\mathbf{A}^{GG_0}\mathbf{F}_i^{G_0}. \qquad (4.51)$$

4.1.3.3 Amonton–Coulomb's Friction Model

In the present problem, other than the normal contact force, frictional contact force also does exist. During momentary impact, the slip angle ε_s is assumed to be zero. Gradually, this momentary impact of the foot end at the initial point of contact $(\mathbf{P}'_{i3})$ is followed by sliding phenomenon of the system or slip (assumed to follow straight path in the XY plane) till the leg lifts off at $\mathbf{P}_{i3}$ for swing (Mahapatra et al. [23]. It is to be remembered that in such scenario the static friction coefficient (μ_s) between the foot and ground steadily changes to dynamic friction coefficient (μ_d). The phenomenon conforms to the Amonton–Coulomb's friction law in this book. Therefore, the tangential forces with respect to frame G are given by

$$\left| F_{ixy}^G \right|_s = \left| \sqrt{\left(F_{ix}^G \right)^2 + \left(F_{iy}^G \right)^2} \right|_s \leq \mu_s F_{iz}^G, \text{ (for static friction, no slip condition)}$$

(4.52)

$$\left| F_{ixy}^G \right|_d = \left| \sqrt{\left(F_{ix}^G \right)^2 + \left(F_{iy}^G \right)^2} \right|_d = \mu_d F_{iz}^G \text{ (for dynamic friction, with slip condition)}$$

(4.53)

where μ_s and μ_d are the coefficients of static and dynamic friction, respectively, which depend on the materials in contact; also F_{ixy}^G is the net tangential force at the contact point. Further, the coefficient of friction (μ_s or μ_d) is the tangent of an angle formed by the normal force (adjacent side) and tangential force (opposite side), that is, $\zeta = \tan^{-1}(F_{ixy}^G / F_{iz}^G)$ where ζ is the friction angle (Fig. 4.3a). In 3D space, a static or dynamical friction cone at a point on any surface is formed by the friction angle (Cheng and Orin [7]; Chen [6]; Takemura et al. [39]; Mahapatra et al. [24]. It is to be further noted that slip will not occur if the net tangential force remains within the static friction cone. However, when slip occurs, the net tangential force is governed by the dynamical friction cone.

To avoid slippage during point contact, the static friction cone constraint is governed by the nonlinear inequality constraint in Eq. (4.52). Further, for smooth execution this nonlinear inequality constraint equation is transformed into a linear set of four inequality constraint equations that forms a friction pyramid within the friction cone. In the present book, the ith leg which is in contact with the terrain is subjected to such a set of four inequality constraint equations and can be expressed in matrix form with respect to G as follows:

$$\mathbf{Q}_i . \mathbf{F}_i^G \geq \mathbf{0}_4$$

(4.54)

Substituting Eq. (4.44) in Eq. (4.54)

$$\mathbf{Q}_i . \mathbf{A}^{GG_0} . \mathbf{F}_i^{G_0} \geq \mathbf{0}_4,$$

(4.55)

where $\mathbf{Q}_i$ is the friction coefficient matrix, as given below.

$$\mathbf{Q}_i = \begin{bmatrix} -1 & 0 & \mu_{\text{eff}} \\ 0 & 0 & \mu_{\text{eff}} \\ 0 & -1 & \mu_{\text{eff}} \\ 0 & 0 & \mu_{\text{eff}} \end{bmatrix}, \tag{4.56}$$

where

$$\mu_{\text{eff}} = \mu_s \Big/ \sqrt{2}, \quad (\text{for static friction}) \tag{4.57}$$

$$= \mu_d \Big/ \sqrt{2} \quad (\text{for dynamic friction}) \tag{4.58}$$

Similarly, Eq. (4.53) corresponds to a dynamic friction cone constraint and comes into picture only during slip. It can be further represented as a set of four linear equality constraint equations for the ith leg in contact with the terrain as follows:

$$\mathbf{Q}_i.\mathbf{F}_i^G = \mathbf{0}_4, \tag{4.59}$$

or

$$\mathbf{Q}_i.\mathbf{A}^{GG_0}.\mathbf{F}_i^{G_0} = \mathbf{0}_4 \tag{4.60}$$

In addition to the friction inequality constraints given by Eq. (4.54) or equality constraint given by Eq. (4.59), another inequality constraint equation must be satisfied by the feet of the robot called normal force inequality constraint. The boundary condition is such that only positive vertical forces (definite foot contact with the terrain) are exerted by the feet on the terrain. Therefore,

$$F_{iz}^{G_0} \geq 0, \tag{4.61}$$

But, Eqs. (4.55) and (4.60) already imply Eq. (4.61). Hence, Eq. (4.61) is a redundant set of equation and may be excluded from the formulation. Further, the torsional frictional constraints (static or dynamic) corresponding to moment components at the point of contact are neglected in the current book unlike the case of soft finger contact, which should satisfy torsional friction constraints (Cheng and Orin [7].

During transition of the state of a system from static to dynamic motion, it is observed that the maximum static friction (stiction) force is always more than the dynamic friction force (developed during sliding motion). This is attributed to the fact that as the relative velocity gradually increases, the friction coefficient gradually decreases from its static to dynamic value. Many of the earlier studies by researchers had disregarded this transition, thereby, neglecting the continuity of friction. In that case, the friction coefficient has a sudden shoot from zero to a finite nonzero value that

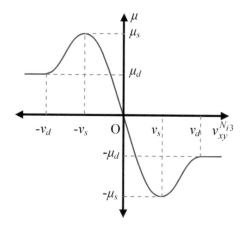

Fig. 4.4 Variation of coefficient of friction (μ) with slip velocity ($V_{xy}^{N_{i3}}$)

never reveals the true phenomenon. To get over such a situation, the present book has admitted this transition phase for continuity of friction. It is analogous to the MSC.ADAMS® Coulomb's friction model. The model is a velocity-based friction model for contacts. Such a friction model can increase the efficiency of compliant contact simulations with smooth transitions as described in Fig. 4.4.

The coefficient of friction is expressed as a function of velocity and is given by (MSC.ADAMS Documentation [28]

$$
\begin{aligned}
\mu(v_{xy}^{N_{i3}}) &= -\mu_d.\mathrm{sgn}(v_{xy}^{N_{i3}}), & \text{for } \left|v_{xy}^{N_{i3}}\right| > v_d \\
&= -f_1(v_{xy}^{N_{i3}}).\mathrm{sgn}(v_{xy}^{N_{i3}}) & \text{for } v_s \leq \left|v_{xy}^{N_{i3}}\right| \leq v_d \\
&= f_2(v_{xy}^{N_{i3}}) & \text{for } -v_s < v_{xy}^{N_{i3}} < v_s
\end{aligned}
\tag{4.62}
$$

The MSC.ADAMS® step functions $f_1\left(v_{xy}^{N_{i3}}\right)$ and $f_2\left(v_{xy}^{N_{i3}}\right)$ are governed by a cubic polynomial [same as Eq. (3.41)] and are given by

$$
f_1\left(v_{xy}^{N_{i3}}\right) = \mathrm{STEP}\left(\left|v_{xy}^{N_{i3}}\right|, v_d, \mu_d, v_s, \mu_s\right),
\tag{4.63}
$$

$$
f_2\left(v_{xy}^{N_{i3}}\right) = \mathrm{STEP}\left(v_{xy}^{N_{i3}}, -v_s, \mu_s, v_s, -\mu_s\right)
\tag{4.64}
$$

where $v_{xy}^{N_{i3}}$ is the slip velocity at contact point; v_s is the stiction transition velocity; v_d is the friction transition velocity; $\mathrm{sgn}(v_{xy}^{N_{i3}})$ is the signum function defined by

$$
\mathrm{sgn}(v_{xy}^{N_{i3}}) = \begin{cases} +1, v_{xy}^{N_{i3}} > 0 \\ 0, v_{xy}^{N_{i3}} = 0 \\ -1, v_{xy}^{N_{i3}} < 0 \end{cases},
\tag{4.65}
$$

It is to be noted that when the value of v_s approaches closer to zero, the model approaches stiction. However, MSC.ADAMS® does not allow $v_s = 0$, so it is impossible to model perfect stiction. The four characteristics (μ_s, μ_d, v_s, v_d) are user-specified. The values of μ_s and μ_d are kept between 0 and 1, such that μ_d is typically lower than μ_s, which is logical, since it is easier to keep a body in motion across a horizontal surface compared to start its motion from rest. Further, it is important to note that v_d is greater than v_s by definition. The coefficients of friction for every slip velocity are calculated which must be multiplied by the normal force $\left(F_{iz}^{G_0} \right)$ to calculate the actual Coulomb friction force.

4.1.4 Static Equilibrium Moment Equation

In the present book, the static equilibrium moment equations for the six-legged robot can be affirmed as the conservation of the moments around the origin G_0 (refer to Fig. 4.5). Therefore,

$$\mathbf{T}_{C_m}^{G_0} + \tilde{\mathbf{r}}_{C_m O}^{G_0} \mathbf{F}_{C_m}^{G_0} + \sum (\mathbf{T}_i^{G_0} + \tilde{\mathbf{r}}_{P_{i3} O}^{G_0} \mathbf{F}_i^{G_0}) = \mathbf{0}_3 \qquad (4.66)$$

where $\tilde{\mathbf{r}}_{P_{i3} O}^{G_0}$ and $\tilde{\mathbf{r}}_{C_m O}^{G_0}$ are the skew symmetric matrix of the location vectors of the foot tip in the support phase and COM of the whole system, respectively, with respect to origin O in frame G_0; $\mathbf{F}_{C_m}^{G_0}$ is the vector of gravitational forces acting on the system denoted by $\begin{bmatrix} 0 & 0 & m_T g \end{bmatrix}^T$, m_T is total mass of the system (refer to Sect. 3.2.8), g is acceleration due to gravity. The term '$\mathbf{T}_i^{G_0} + \tilde{\mathbf{r}}_{P_{i3} O}^{G_0} \mathbf{F}_i^{G_0}$' is the external moment that illustrates the way ground is reacting to the six-legged motion with respect to base-frame origin G_0.

To balance the system, the resultant moment acting on the system about C_m must be zero, that is,

$$\mathbf{T}_{C_m}^{G_0} = 0 \qquad (4.67)$$

Fig. 4.5 Forces and moments with respect to G_0 acting on the COG of a six-legged robot with supported foot tip at any instant (schematic)

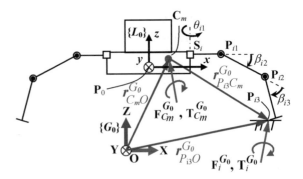

Therefore, Eq. (4.66) reduces to

$$\tilde{\mathbf{r}}^{G_0}_{C_m O} \mathbf{F}^{G_0}_{C_m} + \sum (\mathbf{T}^{G_0}_i + \tilde{\mathbf{r}}^{G_0}_{P_{i3} O} \mathbf{F}^{G_0}_i) = \mathbf{0}_3 \qquad (4.68)$$

4.1.5 Actuator Torque Limits

The resulting joint torques and feet force distributions on the supporting legs also depend on the physical limits of actuator torques that controls the motion of the joints. Therefore, maximum joint torque constraints is given by

$$M_{ij,\min} \leq M^{G_0}_{ij} \leq M_{ij,\max} \text{ for } i = 1-6, \ j = 1-3, \qquad (4.69)$$

where $M^{G_0}_{ij}$ is the torque at the jth joint of ith leg, $M_{ij,\min}$ and $M_{ij,\max}$ are the limiting torques at the jth joint of ith leg. It is to be noted that these limiting torques are usually decided based on the motor specifications.

4.1.6 Optimal Feet Forces' Distributions

As already discussed in Chap. 1, legged systems are superior to traditional vehicles with wheels or tracks with respect to agility, terrain adaptability, and maneuverability. However, from energy consumption point of view, they are still less efficient compared to wheels and tracks. Therefore, in the present state of development, several features need to be optimized.

In the present book on six-legged robots, the inverse dynamic solution to the equilibrium equations (Eq. (4.19)) is not unique, that is, feet forces and moments, joint torques (54 unknowns) have finite number of solutions according to the equation. To determine the best possible solution of the unknowns, the joint torques must satisfy the physical constraints as mentioned in Sects. 4.1.2, 4.1.3, 4.1.4, and 4.1.5. Now, the task is to execute Eq. (4.19) using optimization techniques by introducing an objective function. Literature survey shows that joint torque can be estimated based on the concept, '*power loss in the armature resistance of an electrical motor (actuators) used to rotate a rotary joint is proportional to the square of the joint torque*' (Fukushima and Hirose [10]. Hence, in the present book, minimization of the sum of the squares of the joint torques has been considered as an objective function to make the problem more realistic. The objective function is subjected to equality and inequality constraints, (refer to Sects. 4.1.2, 4.1.3, 4.1.4, and 4.1.5) like force and moment balance equations of the trunk body and payload [Eqs. (4.17) and (4.18)], interactive forces and moments equations [Eqs. (4.44) and (4.51), respectively], friction constraints due to interaction with terrain [Eqs. (4.55), (4.60), and (4.61)], static equilibrium moment equation [Eq. (4.68)], and actuator torque limitation constraints [Eq. (4.69)] on the resulting feet force distribution. It is important to note that whatever method may be adopted to find the force distribution solution, all the constraints associated with the optimization function must be satisfied.

Mathematically, the objective function can be written as follows (Erden and Leblebicioglu [8]; Chen et al. [6]; Li et al. [18]; Mahapatra et al. [24]:

Objective function:

$$S(\mathbf{T}^{G_0}) = \frac{1}{2}\left(\mathbf{M}^{G_0}\right)^{\mathrm{T}} \mathbf{W} \mathbf{M}^{G_0}, \tag{4.70}$$

where $\mathbf{W} \in \mathbb{R}^{18 \times 18}$ is a symmetric positive definite matrix (Erden and Leblebicioglu [8], $\mathbf{M}^{G_0} \in \mathbb{R}^{18}$ is the overall joint torque vector. Equation (4.70) entails quadratic optimization function, which means QP is to be applied. Moreover, QP has some advantages, like, (a) continuous solutions are produced under smooth constraint changes, (b) splitting of the design variables into positive and negative parts is not necessary, (c) more efficient for large problems like the present one, and (d) allows a quadratic optimization rather than linear one.

The function is further expressed in terms of primary variables $\left(\mathbf{F}^{G_0}\right)$, that is, it is rewritten in standard QP form after rearranging the terms (refer to Appendix A.11) as follows:

$$\min_{\mathbf{F}^{G_0}} : \quad S(\mathbf{F}^{G_0}) = \frac{1}{2}\left(\mathbf{F}^{G_0}\right)^{\mathrm{T}} \bar{\mathbf{H}} \mathbf{F}^{G_0} + \mathbf{c}^{\mathrm{T}} \mathbf{F}^{G_0} \tag{4.71}$$

subject to

$$\mathbf{A}_e \cdot \mathbf{F}^{G_0} = \mathbf{B}_e, \tag{4.72}$$

$$\mathbf{A}_u \cdot \mathbf{F}^{G_0} \geq \mathbf{B}_u, \tag{4.73}$$

and

$$\mathbf{F}^{G_0}_{\min} \geq \mathbf{F}^{G_0} \geq \mathbf{F}^{G_0}_{\max} \tag{4.74}$$

where $\bar{\mathbf{H}} \in \mathbb{R}^{18 \times 18}$ is an auxiliary variable called the Hessian matrix. It is a Jacobian square matrix that includes the coefficients of all the quadratic terms of the objective function. $\mathbf{c} \in \mathbb{R}^{18}$ is also an auxiliary variable. $\mathbf{A}_e$, $\mathbf{B}_e$, $\mathbf{A}_u$, and $\mathbf{B}_u$ are the combined matrices as given in Appendix A.11.

Hence, the coupled dynamical system is mathematically expressed as a constrained optimization problem and solved in MATLAB using QP approach to determine the finite ranges of contact forces and moments distributions in all the legs, which are considered to be appropriate according to the criteria described by Eqs. (4.72)–(4.73).

4.1.7 Energy Consumption of a Six-Legged Robot

In multi-legged robots, energy consumption is one of the key factors for its performance. A multi-legged robot maneuvering in different environmental conditions to achieve its goal, consuming minimum energy is one of the most challenging problems. The main reason for the energy consumption in such robots is due to the energy consumed by the actuators attached at the joints of the legs. Further, the consumption of energy by the actuators depends on various gait parameters, vehicle parameters, terrain parameters, etc. Subsequently, the gait parameters include duty factors, stroke, cycle time, swing height, etc. Vehicle parameters include weight of the robot, varying height of the trunk body and velocity (linear/angular) of the trunk body and the swing legs.

Terrain parameters include slope, height of terrain, friction between leg tip and terrain, etc.

In the present work, all the above parameters can be varied for different duty factors. Therefore, for a specific terrain condition, energy consumption depends on all the parameters mentioned above. Finally, to determine the energy consumption at each joint and thereafter by the system, it is necessary to calculate the angular velocities (refer to Sect. 3.2.7) and torques in joints (refer to Sect. 4.1.6).

Energy consumption at each joint is given by

$$P_{ij}(t) = M_{ij}^{G_0}(t).^{G_0}\omega_{ij}(t). \tag{4.75}$$

For the six-legged robots, total energy consumption by the system is given by

$$P_{\text{tot}} = \sum_{i=1}^{6} \sum_{j=1}^{3} P_{ij}(t). \tag{4.76}$$

Therefore, average power consumption by the system is given by

$$P_{av} = \frac{P_{\text{tot}}}{N} = \frac{P_{\text{tot}}}{T/h}, \tag{4.77}$$

where T is the total time taken by the robot to execute the motion, and h is the time step.

Here, it has to be mentioned that for a specified task, the total energy consumption is more important than the average power consumption, since the objective is to minimize the total energy expenditure over a fixed distance. But again, energy is the integral of power consumption over time, which means that minimization of energy results if the average power consumption is minimized. There are possibilities that a particular task is fulfilled in shorter time than expected by consuming higher average energy, so that the total dissipated energy becomes less. In such a scenario, it is more appropriate to compare the energy efficiencies of various types of locomotion of multi-legged robots using energy consumption per weight per travelled distance, commonly termed as, specific energy consumption (Kar et al. [16]; Erden and Leblebicioglu [9] or specific resistance (Marhefka and Orin [25]; Bombled and Verlinden [5]. It is unitless.

Specific energy consumption (E_s) is defined as follows:

$$E_s = \frac{E_D}{m_T g s_T} = \frac{P_{av}}{m_T g v_{t_1}}, \tag{4.78}$$

where
$m_T = m_0 + \sum_{i=1}^{6} \sum_{j=1}^{3} m_{ij}$, (refer to Eq. (3.137)) and
$s_T = n.(ms_0'')$. (refer to Eq. (3.95))
where E_D is the energy required to traverse a total distance of s_T, with a total mass of m_T; P_{av} is the average power consumption; v_{t_1} is the velocity of the robot. Further, Eq. (4.78) is a dimensionless quantity and can be treated as the index of energy efficiency in this study for different types of motion (straight-forward, crab, or turning, etc.).

4.1.8 Stability Measures of Six-Legged Robots

Like, energy consumption, the stability criterion can also be treated as a performance index of legged robots. There are two main types of such criteria, namely (a) static stability criteria and (b) dynamic stability criterion. Whenever, during the locomotion of a legged robot, the static stability criteria are not met, it cannot be characterized as statically stable but is considered to be undergoing the dynamic transient process.

In the present work, both static and dynamic stability criteria have been applied to the six-legged robot to check its stability in different static and dynamic situations. The following sections discuss the statically stable walking based on ESM, NESM, and dynamically stable walking based on DGSM.

4.1.8.1 Statically Stable Walking Based on ESM and NESM

In statically stable walking, the inertial forces acting on the system are neglected. This can be related to the fact of slow speed of statically stable gaits, where gravitational forces are more predominant than motion-dependent forces. Thereafter, the plan is to maintain the vertical projection of the COG of the system within the support polygon (Santos et al. [35] formed by the support points of the feet in contact with the ground. Otherwise, if the projection of the COG is near to the edge of the support polygon, a small momentum of the robot or an external force or uncertainties in the position of COG may bring instabilities in the system and in the worst case, it may cause the robot to topple. This may lead to damage of the system itself and, thereby, failing to locomote as desired. Hence, a measure for stability of the robot like stability margin can be applied in the motion and gait planning, so that, instabilities in the system can be detected and avoided.

The simplest and one of the popular stability criteria is called the static stability margin (SSM). SSM is defined as the minimum distance between the projection of the COG and the edges of the support polygon. The criterion is first proposed by McGhee and Frank [26] based on geometric concepts, which is also independent of COG height. Thereafter, energy-based stability criterion that gives a quantitative measure of the impact energy the robot can withstand without overturning and also considers the height of the COG is proposed by Messuri and Klein [27]. It is called the energy stability margin (ESM) and is defined as the minimum of the potential energy (PE) stability levels required to tumble the robot around the edges of the support polygon. It can be expressed as follows:

$$S_{\mathrm{ESM}} = \min_{l}^{n_g}(m_T g h_l),\qquad(4.79)$$

where l denotes the edge or segment of support polygon considered as the rotation axis, n_g is the number of support/grounded legs, and h_l is the vertical height that gives the measure of energy stability level.

For a six-legged robot, Fig. 4.6 gives a geometrical interpretation for calculation of the vertical height h_l, where points $\mathbf{P}_{13}$ and $\mathbf{P}_{53}$ represent the footholds of two support legs in a tripod gait configuration case. The line segment joining $\mathbf{P}_{13}$ and $\mathbf{P}_{53}$ represents support edge l, more specifically, $l = 1$ of the support polygon $\mathbf{P}_{13}\mathbf{P}_{43}\mathbf{P}_{53}$ for tripod gait configuration. Plane I is a vertical plane, which includes the line $\mathbf{P}_{13}\mathbf{P}_{53}$, that is edge l. $\mathbf{r}_{C_m/l}^{G_0}$ is a vector from line $\mathbf{P}_{13}\mathbf{P}_{53}$ to COG with respect to frame G_0 and is orthogonal to line $\mathbf{P}_{13}\mathbf{P}_{53}$. Unit vector $\hat{z}$ represents the upward vertical direction.

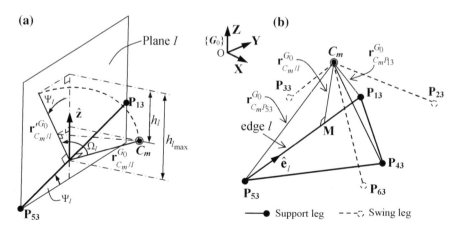

Fig. 4.6 Derivation of energy stability margin **a** Projection of COG of the robot on vertical plane *I*. The line $\mathbf{P}_{13}\mathbf{P}_{53}$ represents support edge *l* of a support polygon, **b** Tripod gait configuration with support polygon $\mathbf{P}_{13}\mathbf{P}_{43}\mathbf{P}_{53}$ and COG of the six-legged walking robot

Now, the vector $\mathbf{r}'^{G_0}_{C_m/l}$ is obtained by rotating vector $\mathbf{r}^{G_0}_{C_m/l}$ about line $\mathbf{P}_{13}\mathbf{P}_{53}$, until it lies in plane I. Also, Ω_l is the angle between $\mathbf{r}^{G_0}_{C_m/l}$ and $\mathbf{r}'^{G_0}_{C_m/l}$, and Ψ_l is the angle between $\mathbf{r}'^{G_0}_{C_m/l}$ and unit vector $\hat{\mathbf{z}}$. Now, for (a) flat plane, $\Psi_l \equiv \alpha_G = 0^0$ (α_G is the sloping angle as defined in Sect. 3.2.1); (b) inclined plane, $\Psi_l \equiv \alpha_G$, and (c) uneven terrain; the value of Ψ_l is determined from the position coordinates of the support legs of the robot.

Hence, the vertical height h_l through which the COG of the system would move when the vector $\mathbf{r}^{G_0}_{C_m/l}$ is rotated about the given edge l to the vertical plane I is given by

$$h_l = \left| \mathbf{r}'^{G_0}_{C_m/l} \right| \cos \Psi_l - \left| \mathbf{r}^{G_0}_{C_m/l} \right| \cos \Omega_l \cos \Psi_l \tag{4.80}$$

Again,

$$\left| \mathbf{r}'^{G_0}_{C_m/l} \right| = \left| \mathbf{r}^{G_0}_{C_m/l} \right| \tag{4.81}$$

Substituting, Eq. (4.81) in Eq. (4.80),

$$h_l = \left| \mathbf{r}^{G_0}_{C_m/l} \right| (1 - \cos \Omega_l) \cos \Psi_l \tag{4.82}$$

Hence, ESM for edge l of the support polygon,

$$S_{ESM} = \min_l^{n_g} \left(m_T g \left| \mathbf{r}^{G_0}_{C_m/l} \right| (1 - \cos \Omega_l) \cos \Psi_l \right) \tag{4.83}$$

S_{ESM} *can be found in a more simplified way, by calculating the maximum height of COG such that*

$$h_{l_{\max}} = \left| \mathbf{r}^{G_0}_{C_m/l} \right| \cos \Psi_l \text{ for } \Omega_l = 90° \tag{4.84}$$

Maximum attainable PE of the system about the edge l of the support polygon is given by

$$(PE)_{\max/l} = m_T g \left| \mathbf{r}^{G_0}_{C_m/l} \right| \cos \Psi_l \tag{4.85}$$

PE of the system is given by

$$(PE)_{sys} = m_T \mathbf{g}^T \mathbf{r}^{G_0}_{C_m O} \tag{4.86}$$

($\mathbf{g}^T$ is the gravitational acceleration vector)
Net potential energy of the system,

$$(PE)_{\text{net}} = (PE)_{\text{max}/l} - (PE)_{\text{sys}}$$

Hence, ESM for edge l of the support polygon,

$$S_{\text{ESM}} = \min_{l}^{n_g}((PE)_{\text{net}}) \tag{4.87}$$

where n_g is the number of grounded legs.

Substituting Eqs. (4.85) and (4.86) in (4.87), we get

$$S_{\text{ESM}} = \min_{l}^{n_g}\left(m_T\left(g\left|\mathbf{r}^{G_0}_{C_m/l}\right|\cos\Psi_l - \mathbf{g}^T\mathbf{r}^{G_0}_{C_m O}\right)\right), \tag{4.88}$$

The vector $\mathbf{r}^{G_0}_{C_m/l}$ which is orthogonal to the edge l is calculated as follows:

$$\mathbf{r}^{G_0}_{C_m/l} = \mathbf{r}^{G_0}_{C_m P_{53}} - \left(\mathbf{r}^{G_0}_{C_m P_{53}} \cdot \hat{\mathbf{e}}_l\right)\hat{\mathbf{e}}_l, \tag{4.89}$$

where

$$\mathbf{r}^{G_0}_{C_m P_{53}} = \mathbf{r}^{G_0}_{C_m O} - \mathbf{r}^{G_0}_{P_{53} O}, \tag{4.90}$$

$$\hat{\mathbf{e}}_l = \left(\mathbf{r}^{G_0}_{P_{13} O} - \mathbf{r}^{G_0}_{P_{53} O}\right) \Big/ \left\|\mathbf{r}^{G_0}_{P_{13} O} - \mathbf{r}^{G_0}_{P_{53} O}\right\|. \tag{4.91}$$

Here, $\hat{\mathbf{e}}_l$ is the unit vector along the edge l with the positive direction from foot $\mathbf{P}_{53}$ toward foot $\mathbf{P}_{13}$ (as shown in Fig. 4.6) to make outward rotation, and $\mathbf{r}^{G_0}_{P_{13} O}$ and $\mathbf{r}^{G_0}_{P_{53} O}$ are the position vectors from the origin of $\mathbf{G}_0$ to the feet $\mathbf{P}_{13}$ and $\mathbf{P}_{53}$ expressed in $\mathbf{G}_0$, respectively.

During the robot's locomotion, the position of all the feet tips (swing and support) with respect to COG of the system (obtained as discussed in Sect. 3.2) and the vector formulation provides a simple efficient method of calculating ESM. It is a more general formulation, which is applicable to variety of terrain conditions. Further, Hirose et al. [14] proposed the normalized energy stability margin (NESM) which is a normalization of the ESM with respect to the robot's weight. It is expressed as follows:

$$S_{\text{NESM}} = \frac{S_{\text{ESM}}}{m_T g} \tag{4.92}$$

In the present book, a stability analysis based on NESM has also been performed in Sect. 4.2.2. Other than these, there are a few other SSMs, which have been developed by researchers worldwide, but are not so widely used.

4.1.8.2 Dynamically Stable Walking Based on DGSM

The static stability measurements, whatsoever, may be very appealing due to simple calculations involved, but such simplicity is achieved at the cost of neglecting the dynamics of the system that may disturb robot's stability. It is seen that the utility of SSM is lost as the legged motions become more dynamic, and hence, a better indicator of stability called dynamic stability measure is required, as discussed in Sect. 2.3. It is to be remembered that dynamic stability in legged robot's can be achieved even if the projection of the center of mass of the robot is not within the support polygon using the dynamic balancing principle. Generally, as discussed in Sect. 2.3, stability margin is evaluated for each supporting edge at any instant. The dynamic stability value for that instant is nothing but the minimum stability margin. In the present book, dynamic stability of six-legged robots has been carried out to determine the chances of tumbling of the walking robot over a support edge by taking into consideration the angular momentum about the edge at any instant. Such stability measure has been termed as dynamic gait stability measure (DGSM). The motivation for such studies has been taken from the work of Koo and Yoon [17] for checking dynamic stability measure for quadruped robots. Several physical quantities could be considered for the DGSM, such as kinetic energy, linear and angular velocities, angular momentum, and moment about a line.

(a)   *Calculation of* $\mathbf{L}^{G_0}$,

Total linear momentum of the system expressed in G_0 is given by

$$\mathbf{L}^{G_0} = \mathbf{A}^{G_0 G}.\mathbf{L}^G, \qquad (4.93)$$

where $\mathbf{L}^G$ is the linear momentum about O of the system expressed in G and is equal to the sum of the linear momentums of the trunk body, payload, and the links of all the legs expressed in G (Refer to Fig. 3.9). It is given by

$$\mathbf{L}^G = \mathbf{L}^G_{C_0 O} + \sum_{i=1}^{6} \sum_{j=1}^{3} \mathbf{L}^G_{C_{ij} O} \qquad (4.94)$$

where $\mathbf{L}^G_{C_0 O}$ is the linear momentum of the trunk body and payload (combined) expressed in frame G; $\mathbf{L}^G_{C_{ij} O}$ is the linear momentum of the jth link in ith leg expressed in frame G.

From linear momentum principle, (Hahn [13] the expression

$$\mathbf{L}^G_{C_0 O} = m_0 \dot{\mathbf{r}}^G_{C_0 O} \qquad (4.95)$$

$$\mathbf{L}^G_{C_{ij} O} = m_{ij} \dot{\mathbf{r}}^G_{C_{ij} O} \qquad (4.96)$$

where $\dot{\mathbf{r}}^{G}_{C_0 O}$ is the time derivative of position vector $\mathbf{r}^{G}_{C_0 O}$ from origin of $\mathbf{G}$ to the COM of the trunk body and payload (combined); $\dot{\mathbf{r}}^{G}_{C_{ij} O}$ is the time derivative of position vector $\mathbf{r}^{G}_{C_{ij} O}$ from origin of $\mathbf{G}$ to the COM of the jth link in ith leg.

Substituting Eqs. (4.95) and (4.96) in Eq. (4.94),

$$\mathbf{L}^{G} = \left(m_0 . \dot{\mathbf{r}}^{G}_{C_0 O} + \sum_{i=1}^{6} \sum_{j=1}^{3} m_{ij} . \dot{\mathbf{r}}^{G}_{C_{ij} O} \right) \tag{4.97}$$

(b) *Calculation of* $\mathbf{H}^{G_0}$

Total angular momentum of the system expressed in $\mathbf{G}_0$ is given by

$$\mathbf{H}^{G_0} = \mathbf{A}^{G_0 G} . \mathbf{H}^{G} \tag{4.98}$$

where $\mathbf{H}^{G}$ is the total angular momentum about $\mathbf{O}$ of the system expressed in $\mathbf{G}$ and is equal to the sum of the angular momentums of the trunk body, payload, and the links of all the legs expressed in $\mathbf{G}$. It is given by,

$$\mathbf{H}^{G} = \mathbf{H}^{G}_{0} + \sum_{i=1}^{6} \sum_{j=1}^{3} \mathbf{H}^{G}_{ij} \tag{4.99}$$

where $\mathbf{H}^{G}_{0}$ is the angular momentum of the trunk body and payload (combined) about $\mathbf{O}$ expressed in frame $\mathbf{G}$; $\mathbf{H}^{G}_{ij}$ is the angular momentum of the jth link in ith leg about $\mathbf{O}$ expressed in frame $\mathbf{G}$.

For trunk body and payload,

$$\mathbf{H}^{G}_{0} = \mathbf{r}^{G}_{C_0 O} \times \mathbf{L}^{G}_{C_0 O} + \mathbf{r}^{G}_{P_0 O} \times \mathbf{L}^{G}_{C_0 P_0} + \mathbf{H}^{G}_{P_0} \tag{4.100}$$

[for details refer to (Hahn [13]]
where

$$\mathbf{L}^{G}_{C_0 P_0} = m_0 . \dot{\mathbf{r}}^{G}_{C_0 P_0} \tag{4.101}$$

$$\mathbf{H}^{G}_{P_0} = \mathbf{A}^{GL_0} . \mathbf{J}^{L_0 G}_{P_0} . \boldsymbol{\omega}_0 \tag{4.102}$$

For link ij,

$$\mathbf{H}^{G}_{ij} = \mathbf{r}^{G}_{C_{ij} O} \times \mathbf{L}^{G}_{C_{ij} O} + \mathbf{r}^{G}_{P_{ij} O} \times \mathbf{L}^{G}_{C_{ij} P_{ij}} + \mathbf{H}^{G}_{P_{ij}} \tag{4.103}$$

where

$$\mathbf{L}^{G}_{C_{ij} P_{ij}} = m_{ij} . \dot{\mathbf{r}}^{G}_{C_{ij} P_{ij}} \tag{4.104}$$

$$\mathbf{H}^G_{P_{ij}} = \mathbf{A}^{GL'_{ij}} . \mathbf{J}^{L'_{ij}}_{P_{ij}} . ^G\boldsymbol{\omega}_{ij} \tag{4.105}$$

Here, $\mathbf{A}^{GL_0}$ and $\mathbf{A}^{GL'_{ij}}$ are the transformation matrices, and the details of which are obtained from Appendix A.3; $\mathbf{J}^{L_0}_{P_0}$ is the mass moment of inertia matrix of the trunk body with respect to the point $\mathbf{P}_0$, represented in the local (body-fixed) frame L_0; $\mathbf{J}^{L'_{ij}}_{P_{ij}}$ is the mass moment of inertia matrix of the trunk body with respect to the point $\mathbf{P}_{ij}$, represented in the local (body-fixed) frame L'_{ij}. For details of mass moment of inertia, refer to Appendix A.8.

Substitution of the values of equation in Eq. (4.99) leads to (Mahapatra et al. [24]

$$\begin{aligned}
\mathbf{H}^G &= \mathbf{r}^G_{C_0 O} \times \mathbf{L}^G_{C_0 O} + \mathbf{r}^G_{P_0 O} \times \mathbf{L}^G_{C_0 P_0} + \mathbf{H}^G_{P_0} \\
&\quad + \sum_{i=1}^{6} \sum_{j=1}^{3} \left(\mathbf{r}^G_{C_{ij} O} \times \mathbf{L}^G_{C_{ij} O} + \mathbf{r}^G_{P_{ij} O} \times \mathbf{L}^G_{C_{ij} P_{ij}} + \mathbf{H}^G_{P_{ij}} \right) \\
&= \left(\mathbf{r}^G_{C_0 O} \times m_0 . \dot{\mathbf{r}}^G_{C_0 O} + \mathbf{r}^G_{P_0 O} \times m_0 . \dot{\mathbf{r}}^G_{C_0 P_0} \right) \\
&\quad + \sum_{i=1}^{6} \sum_{j=1}^{3} \left(\mathbf{r}^G_{C_{ij} O} \times m_{ij} . \dot{\mathbf{r}}^G_{C_{ij} O} + \mathbf{r}^G_{P_{ij} O} \times m_{ij} . \dot{\mathbf{r}}^G_{C_{ij} P_{ij}} \right) \\
&\quad + \mathbf{A}^{GL_0} . \left(\mathbf{J}^{L_0}_{P_0} . ^G\boldsymbol{\omega}_0 + \sum_{i=1}^{6} \sum_{j=1}^{3} \mathbf{A}^{L_0 L'_{ij}} . \mathbf{J}^{L'_{ij}}_{P_{ij}} . ^G\boldsymbol{\omega}_{ij} \right)
\end{aligned} \tag{4.106}$$

(c) *Angular momentum with respect to a supporting edge on the ground*

To determine the angular momentum about supporting edge l (refer to Fig. 4.6), the moment of the system with respect to any ground point (say $\mathbf{P}_{13}$) has to be calculated. Hence, angular momentum of the system about a point $\mathbf{P}_{13}$ on the ground is given by

$$\mathbf{H}^{G_0}_{P_{13}} = \mathbf{r}^{G_0}_{P_{13} O} \times \mathbf{L}^{G_0} + \mathbf{H}^{G_0} \tag{4.107}$$

where $\mathbf{r}^{G_0}_{P_{13} O}$ is the relative position vector $\mathbf{r}_{P_{13} O}$ from point $\mathbf{P}_{13}$ to the origin $\mathbf{O}$ expressed in G_0.

Again, the rotation of the system about the edge l is given by the angular momentum vector of Eq. (4.107) projected to the edge l. Mathematically,

$$H^{G_0}_l = \mathbf{H}^{G_0}_{P_{13}} . \hat{\mathbf{e}}_l = \left(\mathbf{r}^{G_0}_{P_{13} O} \times \mathbf{L}^{G_0} + \mathbf{H}^{G_0} \right) . \hat{\mathbf{e}}_l \tag{4.108}$$

(d) *Reference angular momentum about the supporting edge on the ground*

According to conservation of energy,

$$(\text{KE})_{\text{net}} = (\text{PE})_{\text{net}}$$

$$\left| -\tfrac{1}{2} m_T v_{ref}^2 \right| = m_T \left(g \left| \mathbf{r}_{C_m/l}^{G_0} \right| \cos \Psi_l - \mathbf{g}^{\mathrm{T}} \mathbf{r}_{C_m O}^{G_0} \right)$$

$$v_{ref} = \sqrt{2 \left(g \left| \mathbf{r}_{C_m/l}^{G_0} \right| \cos \Psi_l - \mathbf{g}^{\mathrm{T}} \mathbf{r}_{C_m O}^{G_0} \right)} \tag{4.109}$$

where v_{ref} is the magnitude of the reference velocity, and $\mathbf{r}_{C_m O}^{G_0}$ is the position of the center of mass of the system obtained from Eq. (3.138).

The reference velocity vector $\mathbf{v}_{ref}$ is given by

$$\mathbf{v}_{ref} = v_{ref} . \hat{\mathbf{e}}_{ref} \tag{4.110}$$

where $\hat{\mathbf{e}}_{ref}$ is the direction of the reference velocity and is outer normal to the plane formed by the locations of tip points $\mathbf{P}_{13}$ and $\mathbf{P}_{53}$ and $\mathbf{C}_m$ of the system. Mathematically, it is expressed by

(i) $\hat{\mathbf{e}}_{ref} = \dfrac{\mathbf{r}_{P_{53}C_m}^{G_0} \times \mathbf{r}_{P_{13}C_m}^{G_0}}{\left| \mathbf{r}_{P_{53}C_m}^{G_0} \times \mathbf{r}_{P_{13}C_m}^{G_0} \right|}$, If $Z-$ component of $\mathbf{r}_{P_{53}C_m} \times \mathbf{r}_{P_{13}C_m}$ is positive

$$\tag{4.111}$$

(ii) $\hat{\mathbf{e}}_{ref} = -\dfrac{\mathbf{r}_{P_{53}C_m}^{G_0} \times \mathbf{r}_{P_{13}C_m}^{G_0}}{\left| \mathbf{r}_{P_{53}C_m}^{G_0} \times \mathbf{r}_{P_{13}C_m}^{G_0} \right|}$, If $Z-$ component of $\mathbf{r}_{P_{53}C_m} \times \mathbf{r}_{P_{13}C_m}$ is negative

$$\tag{4.112}$$

The statement in the right-hand side of Eqs. (4.111) and (4.112) designates whether the COG has passed over the edge or not. If Z-component of $\mathbf{r}_{P_{53}C_m}^{G_0} \times \mathbf{r}_{P_{13}C_m}^{G_0}$ is positive, COG has not tipped over the edge yet. Therefore, reference angular momentum about a supporting edge l may be defined as the minimum angular momentum needed to tip over the edge, assuming the robotic system as a simple inverted pendulum. It is computed as follows:

$$H_l^{\mathrm{ref}} = \left(\mathbf{r}_{C_m/l}^{G_0} \times m_T \mathbf{v}_{ref} \right) \hat{\mathbf{e}}_l \tag{4.113}$$

Equations (4.89), (4.91), and (4.110) are substituted in Eq. (4.113) to obtain the reference angular momentum about edge l.

(e) *Dynamic gait stability measure (DGSM)*

Gait stability value for edge l is defined as follows:

$$S_H^l = H_l^{\mathrm{ref}} - H_l^{G_0}, \tag{4.114}$$

where the value of S_H^l should always be positive; otherwise, the system will topple due to dynamical instability. The value of S_H^l with respect to all the edges of the support polygon at any instant is calculated. Therefore, DGSM which is nothing but the minimum value among the stability values S_H^l is given by the following expression:

$$\text{DGSM} = \min\left\{S_H^l, l = 1, 2, \ldots n_l\right\} \tag{4.115}$$

4.2 Numerical Illustrations

The performance of the developed dynamic model has been tested using computer simulation. The additional input parameters over and above that are mentioned in Sect. 3.3 are as follows:

Additional inputs

- Stiffness (K)
- Force exponent (e)
- Damping $(C_{\max})$
- Penetration depth (p)
- Static coefficient of friction (μ_s)
- Dynamic coefficient of friction (μ_d)
- Stiction transition velocity (v_s)
- Friction transition velocity (v_d)
- Mass of the trunk body (m_B)
- Mass of payload (m_L)
- Mass of the links (m_{ij})
- Mass moment of inertias $(J_{xx}, J_{yy}, J_{zz}, J_{xy}, J_{yz}, \text{and } J_{zx})$

Simulation Outputs

- Feet forces ($[F_{ix}^{G_0} \ F_{iy}^{G_0} \ F_{iz}^{G_0}]^\text{T}$)
- Joint torques ($[M_{i1}^{G_0} \ M_{i2}^{G_0} \ M_{i3}^{G_0}]^\text{T}$)
- Feet-tip moments $[T_{ix}^{G_0} \ T_{iy}^{G_0} \ T_{iz}^{G_0}]^\text{T}$
- Energy consumption (P_{av})
- NESM
- DGSM

A flowchart shown in Fig. 4.7 illustrates the steps to be followed to carry out the inverse dynamic analysis.

4.2.1 Study of Optimal Feet Forces' Distribution

In the present work, numerical simulations have been carried out under a variety of terrain conditions and gait strategies to check the performance of the proposed

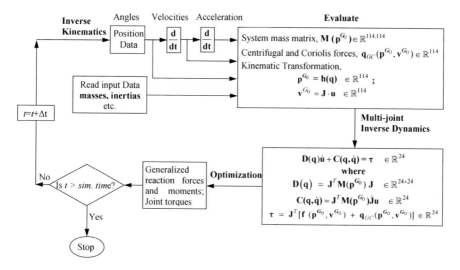

Fig. 4.7 Flowchart of computational algorithm for the inverse dynamic analysis of the six-legged robot negotiating various terrains

Table 4.2 Contact parameters

Stiffness (K)	10×10^8 N/m
Force exponent (e)	2.2
Damping (C_{max})	10×10^6 N/m/s
Penetration depth (p)	1.0×10^{-5} m
Static coefficient of friction (μ_s)	0.3
Dynamic coefficient of friction (μ_d)	0.1
Stiction transition velocity (v_s)	0.01 m/s
Friction transition velocity (v_d)	0.1 m/s

algorithm as discussed in Sect. 4.1. The physical parameters' values of the robot model are as described in Table 3.1. The contact parameters' (refer to Table 4.2) values supplied as used inputs to the proposed model are basically assumed to be default input values used by MSC.ADAMS® to solve contact problems. It is to be noted by varying the parameters (as shown in Table 4.2), a variety of walking and running surfaces can be simulated.

4.2.1.1 Case Study 1: Robot Motion in an Uneven Terrain with Straight-Forward Motion (DF = 1/2)

An attempt has been made in the present work to carry out the dynamic analysis of the case study included in Sect. 3.3.1. The kinematic analysis of the system is performed with predefined input parameters using MATLAB solver as already discussed in

Sect. 3.3.1. Here, multi-body dynamic analysis of the system has been carried out with the motion inputs obtained from the study of kinematics. The simulation study is conducted in MATLAB for three duty cycles ($n = 3$) with time step $h = 0.05$ s and trunk body stroke length (s_0) of 0.150 m.

The elapsed time for three duty cycles is computed as 6.95 s. The computed kinematic motion parameters (displacement, velocity, acceleration, etc.), based on the motion and gait planning algorithms, are provided as necessary inputs to the inverse dynamics model with compliant contact. Besides those parameters, the physical and contact parameters defined in Tables 3.1 and 4.2 are also given as the inputs. The study is carried out with the assumption that there is no slippage of the leg pad on the surface. Additionally, the actuator torque is limited to ± 6 Nm for calculating the optimal joint torques, feet forces, etc. The optimization algorithm used is *interior-point-convex quadprog* that satisfies the boundary conditions corresponding to the objective function for each iteration.

Figure 4.8 shows the optimal distribution of normal feet forces ($F_{iz}^{G_0}$) in all the legs of the six-legged robot. It is observed that with wave gait strategy (DF = 1/2), the support phase and swing phase times are equal. It followed a wave sequence like first half cycle: legs 1-4-5 are in support phase and legs 2-3-6 are in swing phase; second half cycle: legs 2-3-6 are in support phase and legs 1-4-5 are in swing phase. The sequence is followed for all the three cycles. Investigation reveals that the patterns of the normal feet forces ($F_{iz}^{G_0}$) in legs 1-2, 3-4, and 5-6 are similar, although the patterns are out of phase by $180°$ and a slight variation in the magnitude of feet forces' distribution is observed. Similar patterns is attributed to the fact that wave gaits are regular and symmetric, with the left and right legs (i.e., legs 1-2, 3-4, 5-6) are in phase of a half cycle. The variation in magnitude is attributed to the possibility of the robot locomoting on an uneven terrain in the present problem. The graphs also show that as the leg collided with the ground, high shoot out in forces is observed, which is attributed to the momentarily impact of the leg tip with the ground. This is considered as realistic in the present work. The zero magnitude straight line in the graphs (refer to Fig. 4.8) means no contact of the leg tip with the ground, which is due to the leg being in swing phase. Further investigation reveals that as the robot executes motion, the magnitude of normal feet forces on the front legs increases while that of the rear legs decreases till the start of a new cycle. This is due to the fact that as the leg moves forward in a straight-forward motion, the entire COG of the system shifts forward leading to more load distribution in front legs compared to that in the rear legs (refer to Fig. 4.8a, b, e, and f). Also, it is interesting to note that the normal feet forces' variation in the middle legs is less and magnitude is somewhat constant (refer to Fig. 4.8c and d), which is due to the close location of the foot holds to the COG of the system during the support phase. Further, the average vertical forces are more in the middle legs compared to that of the front and rear legs. The distribution patterns are found to be similar to that given by Santos et al. [34].

Figure 4.9 shows that at any instant of time the summation of all the normal feet forces with respect to G_0 balanced the weight of the six-legged robot with payload (i.e., 65.7 N) except during impact. A momentary impulsive force is generated during impact at the start of support phase, such that Fig. 3.9a and b displayed the horizontal

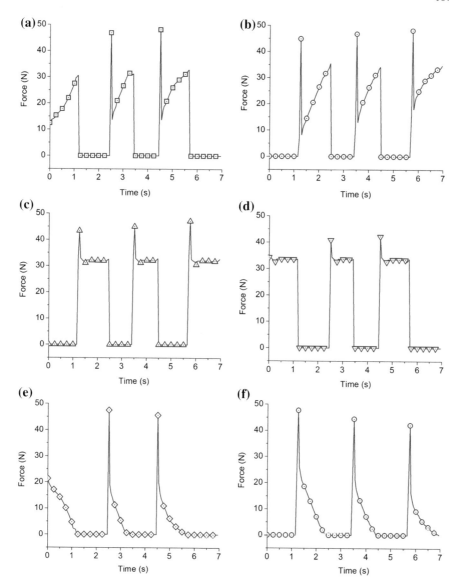

Fig. 4.8 Normal contact force distribution on the leg tips of the six-legged robot with respect to frame G_0 while moving in an uneven terrain with tripod gait **a** Leg1, **b** Leg2, **c** Leg3, **d** Leg4, **e** Leg5, **f** Leg6

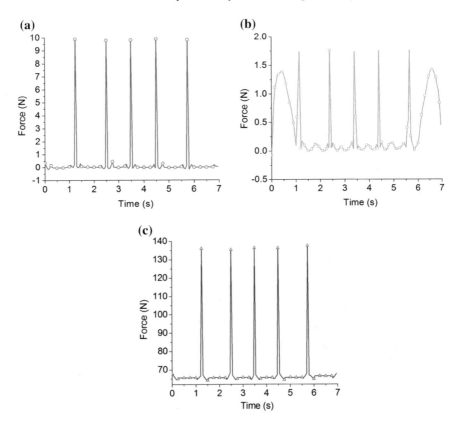

Fig. 4.9 Summation of forces distribution with respect to frame G_0 **a** $\sum F_{ix}$, **b** $\sum F_{iy}$, **c** $\sum F_{iz}$, where $i = 1$–6 for DF = 1/2 with straight-forward motion on an uneven terrain

components of forces ($F_{ix}^{G_0}$ and $F_{iy}^{G_0}$) which did exist, although their net magnitude is not significant enough (approximately in the range of 0.1–0.2 N if momentary impact force is neglected). Similarly, a momentarily shoot out peak in the normal reaction foot force $F_{iz}^{G_0}$ is also observed in Fig. 4.8c.

Figure 4.10 shows the torque distributions in various joints (joint $i1$, joint $i2$, and joint $i3$) of all the legs ($i = 1$–6). Some important facts are noted from the plots, like torques during the swing phases are low and subjected to less variations in comparison with that of the torques acting on the joints during support phase. Moreover, during support phase, the maximum torque is experienced by middle joint (joint $i2$). For the front legs (leg 1 and 2), the joint torque increased, while in the rear legs (leg 5 and 6) it is seen to decrease. This is accredited to the fact that during straight-forward motion, the body moves forward, leading COG of the system to move forward. The forward movement of the COG indicates an increase in front loading and decrease in rear loading in the system. Again, the middle legs are subjected to less variations due to closeness of the support legs toward the COG. The most striking fact to be discussed is that of the effect of momentarily impact on the joint torque, when the leg collides

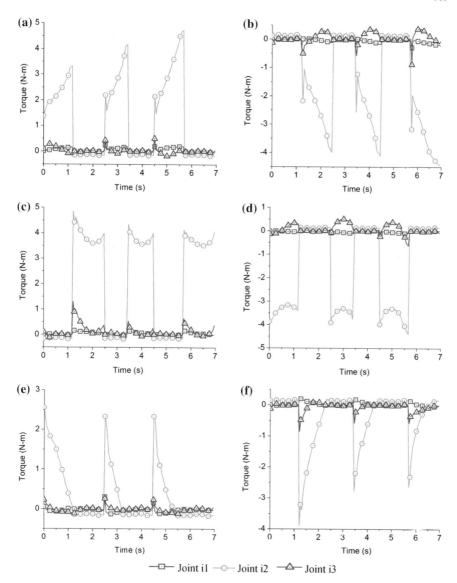

Fig. 4.10 Torque distribution in various joints of the six-legged robot while traversing an uneven terrain with tripod gait and straight-forward motion **a** Leg1, **b** Leg2, **c** Leg3, **d** Leg4, **e** Leg5, **f** Leg6

with the ground at the start of support phase. It is seen that there is momentary shoot out peak in the magnitude of joint torque of its own leg while the impact is seen to have considerable effect on the other legs (small spikes) due to coupling effects.

The variations of instantaneous power (P_{in}) consumption throughout a locomotive cycle of the six-legged robot are shown in Fig. 4.11. There is momentary change

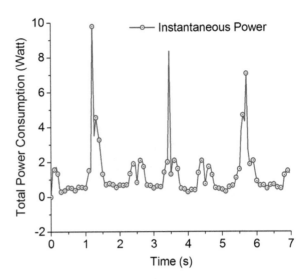

Fig. 4.11 Variation of total power consumption by the joints of all the legs during straight-forward motion with tripod gait on an uneven terrain

(increase or decrease) in the value of instantaneous power consumption at intervals, when the support and swing legs change stance. The average power (P_{av}) consumed by the system at any instant is computed to be 1.27 W neglecting friction losses at the joints.

4.2.1.2 Case Study 2: Crab Motion of the Robot on a Banked Surface (DF = 3/4)

The present case study deals with the dynamic analysis of a realistic six-legged robot whose kinematic analysis has been carried out in Sect. 3.3.2. As already discussed, kinematic analysis of the system has been performed with the help of computer simulations in MATLAB solver by predefining the input parameters. In the current section, multi-body dynamic analysis of the system has also been carried out through computer simulations in MATLAB, to prove the efficacy of the developed algorithm. Relevant kinematic motion inputs (like displacement, velocity, acceleration, etc., computed in Sect. 3.3.2) along with other physical inputs mentioned in and Table 4.2 are used to solve the present problem. The study has been carried out with actuator torque limits of ±6 Nm and without slippage for calculation of optimal joint torques, feet forces, power consumption, etc. The optimization algorithm selected is *interior-point-convex quadprog* that satisfies the boundary conditions corresponding to the objective function for each iteration. Computer simulations are run for three cycles, with the body stroke (s_0) of 0.027 m and time step size of 0.01 s as previously mentioned. The total simulation time is 5.9 s with time period of first, second, and third cycles as 2.05 s, 1.80 s, and 2.05 s, respectively, as computed using Eq. (3.92).

Figure 4.12 shows the optimal distribution of normal feet forces obtained through computer simulations. It is seen that for each cycle with DF = 3/4, the robot's legs

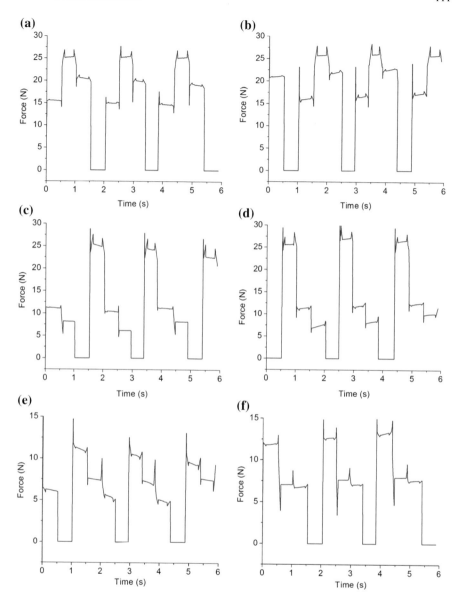

Fig. 4.12 Normal contact force distribution on the leg tips of the six-legged robot with respect to frame G_0 while moving on a banking surface with crab motion and DF = 3/4 **a** Leg 1, **b** Leg 2, **c** Leg 3, **d** Leg 4, **e** Leg 5, **f** Leg6

follow a wave pattern such that the legs in support phases are sequenced by (1) legs 1-2-3-5-6 (2) legs 1-3-4-6 (3) legs 1-2-4-5-6 (4) legs 2-3-4-5. Further, it is observed that although the legs 1-2, 3-4, and 5-6 are out of phase by 180°, their normal feet forces follow a similar pattern. Also, it is seen that impact forces are generated when the legs collide with the terrain. These impact forces are nothing but momentary peaks (very short duration, i.e., in the range of milliseconds) as shown in Fig. 4.12.

It is to be noted that the momentary impact of forces on the leg that is landing has effects on the force distribution in the other legs in support phase. This is caused due to coupling effects in the system. Again, it is seen that for any cycle, the normal force distribution ($F_{iz}^{G_0}$) in the front legs (leg 1 and 2) is more compared to that in the rear legs (leg 5 and 6). This is attributed to the fact that as the robot's leg moves forward and to the right (crab angle = 70°), along with the trunk body, the COG of the system also moves in that direction. This leads to more force distribution in the front legs compared to that of the rear legs. It is also noticeable in Fig. 4.12 that when legs 1 and 6 are in swing phase, that is, legs 2-3-4-5 are in the support phase, force distribution in legs 2 and 3 is more compared to legs 4 and 5. Again, both in the legs 2 and leg 3, force distribution is almost equal and so is the case with leg 4 and leg 5. A similar case has been observed when leg 2 is in swing phase (refer to Fig. 4.12.) and so on.

Again from Fig. 4.13, it is seen that at any instant of time, the sum of all the normal feet forces with respect to frame G_0 balanced the weight of the six-legged robot with payload (i.e., 65.7 N) except during impact (refer to Fig. 4.13c). Further, it is observed from the graph that at the instant the foot comes in contact with the ground (i.e., start of support phase), a momentary impulsive force is generated whose effects are also visualized in the graphs showing horizontal components of forces, that is, $F_{ix}^{G_0}$ and $F_{iy}^{G_0}$, (Fig. 4.13a and b) as well in the normal reaction force, $F_{iz}^{G_0}$ (Fig. 3.13c). The net magnitude of forces in X- and Y-direction of frame G_0 is very small and in approximate ranges of ±0.5 N and −0.15 N to +0.05 N, respectively.

Figure 4.14 shows the torque distribution in the various joints of the legs for three duty cycles with DF = 3/4. It is seen that during swing phase, the variations of torque in various joints are less compared to that during support phase. Also, it is observed that the magnitude of torque in the first and last joints of each leg is much less in comparison with that on the middle joints. In addition to that, torque distribution in the middle joint of the rear legs is less compared to that in the middle joint of the front legs. In addition to that, torque in middle joint of the front legs is more compared to that of the rear legs, which is attributed to the different values of moment arms during crab motion from left to right (positive X-axis) at the given crab angle of 70°. Further, it is interesting to note that the momentary impact forces are generated as soon as the legs collide with the ground during support phase. The impact forces are seen to significantly affect the torques in the joints of its own legs while having considerable effect on the joints of other legs (small peaks) due to the coupling effects. Refer to the close view of 'A' and 'B' in the Fig. 4.14.

The instantaneous power consumption (P_{in}) by the robot is as shown in Fig. 4.15. It is found that at times when the swing and support legs change stance peaks are generated that have either increasing or decreasing trend. The average power (P_{av})

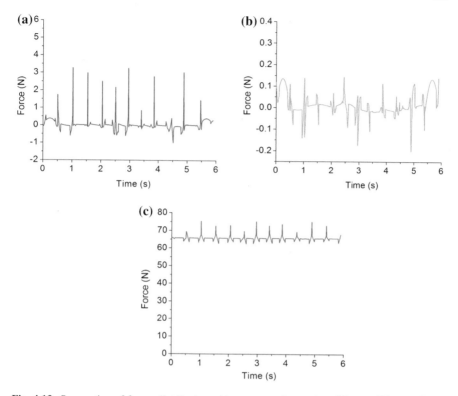

Fig. 4.13 Summation of forces distribution with respect to frame G_0 **a** $\sum F_{ix}$, **b** $\sum F_{iy}$, **c** $\sum F_{iz}$, where $i = 1–6$ for DF = 3/4 and crab motion on a banking surface

consumed by the system (neglecting friction losses at the joints) in three cycles is 1.16 W (approximately).

4.2.2 Study of Performance Indices—Power Consumption and Stability Measure

Energy consumption and stability are two of the important criteria that have been selected as useful indices in the present study to evaluate the static and dynamic walk of the six-legged robot during wave gait strategy. Here, locomotion of the robot has been considered on various terrains having straight-forward, crab, and turning motion capabilities and using wave gait strategies with DF = 1/2, 2/3, and 3/4. Further, computer simulations are run for studying four important walking parameters that decide the power consumption and stability of the robot during locomotion with different gait strategies as mentioned above. The parameters are, namely (a) velocity of the trunk body ($\dot{y}^G_{P_0O}\big|_{t=1}$ for straight-forward or $\dot{x}^G_{P_0O}\big|_{t=t_1}$ for crab or $\dot{\theta}_0\big|_{t=t_1}$ for

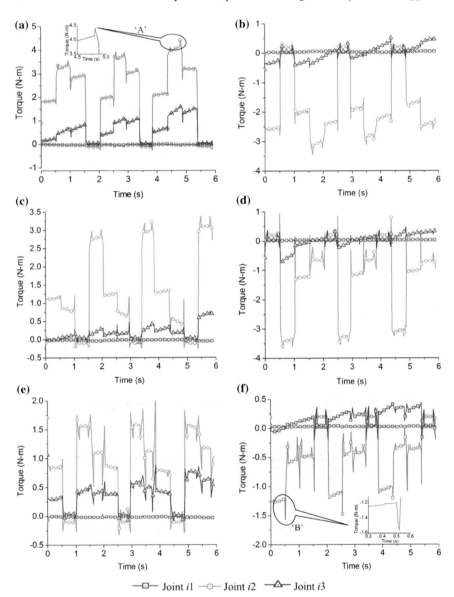

Fig. 4.14 Torque distribution in various joints of the six-legged robot while moving on a banked surface (banking angle 30°) with DF = 3/4 and crab motion **a** Leg1, **b** Leg2, **c** Leg3, **d** Leg4, **e** Leg5, **f** Leg6

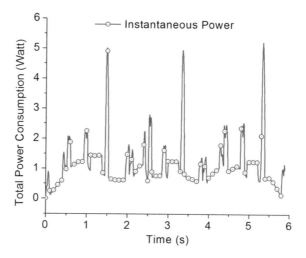

Fig. 4.15 Variation of total power consumption by the joints of all the legs during crab motion with DF = 3/4 on a banked surface

turning), (b) body stroke (s_0/s_0^c), (c) height of the trunk body $(z_{P_0O}^G)$, and (d) lateral or radial offset of leg (l_i). In addition to this, crab angle (θ_c) is also considered as one of the variable parameters in the present study. The energy consumption of the system is expressed in terms of outputs like (i) average power consumption (P_{av}), (ii) specific energy consumption (E_s), whereas the stability of the system is represented in terms of outputs like (iii) net energy stability margin (NESM), and (iv) dynamic gait stability measure (DGSM). The formulation of the above outputs may be referred to Sects. 4.1.7 and 4.1.8.

The kinematic analysis of the system has been carried out using the inputs as mentioned below (notations have their usual meaning).

I. *Straight-forward motion*: For the robot, initial position of the trunk body is given by, $\mathbf{p}_0^G = \{0, \quad 0.45, \quad z_{P_0O}^G, 0, \quad 0, \quad 0\}^T$, $\dot{x}_{P_0O}^G|_{t=t_1} = 0$, $\dot{z}_{P_0O}^G|_{t=t_1} = 0$, $\dot{\alpha}_0 = \dot{\beta}_0 = \dot{\theta}_0 = 0$. For the terrain, $\mathbf{\eta}_G = (30°, 0, 0)^T$, $Hm_{min} = 0.015$ m, $h'_{in} = 0$, $\Delta h = 0.002$ mm. Moreover, the initial joint angles $(\theta_{i1}, \beta_{i2}, \beta_{i3})$ of the legs are also calculated from inverse kinematics formulation and given as inputs in the simulation. The simulations are run for three duty cycles $(n = 3)$ with DF = 1/2, 2/3, and 3/4 for various parameters under study, likely $\dot{y}_{P_0O}^G|_{t=t_1}$, s_0, $z_{P_0O}^G$ and l_i, respectively, in MATLAB. Unless otherwise mentioned, the values of the parameters were $\dot{y}_{P_0O}^G|_{t=t_1} = 0.1$ m/s; $s_0 = 0.15$ m; $z_{P_0O}^G = 0.15$ m; and $l_i = 0.22$ m.

II. *Crab motion*: For the robot, the initial values are expressed as $\mathbf{p}_0^G = \{0, \quad 0.45, \quad z_{P_0O}^G, \quad 0, \quad 0, \quad 0\}^T$, $\dot{z}_{P_0O}^G|_{t=t_1} = 0$, $\dot{\alpha}_0 = +0.01$ rad/s, $\dot{\beta} = +0.02$ rad/s. For the terrain, $\mathbf{\eta}_G = (0, 30°, 0)^T$, $Hm_{min} = 0.015$ m, $h'_{in} = 0$, $\Delta h = 0.002$ m. The initial joint angles $(\theta_{i1}, \beta_{i2}, \beta_{i3})$ of the legs are calculated from inverse kinematics formulation and given in the simulation as inputs. Like straight-forward motion, the simulations are run for three duty cycles $(n = 3)$ with DF = 1/2, 2/3, and 3/4 for various parameters to be studied, like $\dot{x}_{P_0O}^G|_{t=t_1}$,

s_0, $z_{P_0O}^G$, l_i and θ_c, respectively, in MATLAB. It is to be noted that the value of $\dot{y}_{P_0O}^G\big|_{t=t_1}$ is obtained using the Eq. (3.66). Unless otherwise mentioned, the values of the parameters are $\dot{x}_{P_0O}^G\big|_{t=t_1} = 0.04$ m/s; $s_0 = 0.03$ m; $z_{P_0O}^G = 0.15$ m; $l_i = 0.22$ m; and $\theta_c = 80°$.

III. *Turning motion*: For the robot, the initial values of $\mathbf{p}_0^G$ are computed by integrating the Eqs. (3.54) and (3.44). Also, the following values are considered: $\theta_0\big|_{t=t_0} = 30°$, $\rho_0 = 1.5$ m, $\dot{\alpha}_0 = +0.01$ rad/s, $\dot{\beta} = +0.02$ rad/s. For the terrain, the following numerical values are set: $\boldsymbol{\eta}_G = (0, 30°, 0)^T$, $Hm_{in} = 0.005$ m, $h'_{in} = 0$, $\Delta h = 0.002$ m,. The initial joint angles (θ_{i1}, β_{i2}, β_{i3}) of the legs that vary with the different types of variable parameters and gait strategies are also given as inputs. Like straight-forward and crab motions, the simulations are run for three duty cycles ($n = 3$) with DF = 1/2, 2/3, and 3/4 for various parameters, like $\dot{\theta}_0\big|_{t=t_1}$, s_0^c, $z_{P_0O}^G$ and l_i in MATLAB. Unless otherwise mentioned, the values of the parameters are: $\dot{\theta}_0\big|_{t=t_1} = 0.1$ rad/s; $s_0^c = 0.12$ m; $z_{P_0O}^G = 0.15$ m; and $l_i = 0.22$ m.

Further, dynamic analysis has also been carried out with the assumption that the foot tip touches the ground with impact and leg pad does not slip on the surface. The physical and contact parameters related to the six-legged robot given in Tables 3.1 and 4.2, respectively, are also supplied as the inputs. Additionally, the actuator torque is limited to ±6 N-m for calculation of optimal feet forces, joint torques, etc., and *interior-point-convex quadprog* has been chosen as the optimization algorithm such that at each iteration, the boundary conditions corresponding to the objective function are satisfied. Energy consumption and stability of the robot are analyzed by varying the abovementioned gait parameters, which are discussed, in detail, below.

4.2.2.1 Effect of Trunk Body Velocity on Energy Consumption and Stability

The effect of trunk body velocity on energy consumption and stability in the six-legged robots is quite significant. In this study, the trunk body velocity is an independent variable and plotted along X-axis, while the energy consumption and stability measure are dependent variables and plotted along Y-axis. The results are shown in Figs. 4.16 and 4.17, respectively. It is observed from Fig. 4.16a, c, and e that for a particular value of the duty factor, average power consumption (P_{av}) increased with velocity for all the three types of motion (straight-forward, crab, and turning). However, total power consumption will increase with the decrease in trunk body velocity (linear/angular), since to travel the same amount of distance; the robot would have to spend more time leading to more consumption of energy. Therefore, the energy consumption is quantified based on specific energy consumption as discussed in Sect. 4.1.7. Specific energy consumption (E_s) is found to decrease with increase in velocity (linear/angular) as shown in Fig. 4.16b, d, and f. Moreover, for a particular velocity, P_{av} and E_s are seen to increase with the increase in duty factor (DF). This is because the time period of a locomotion cycle decreases with the increase in duty

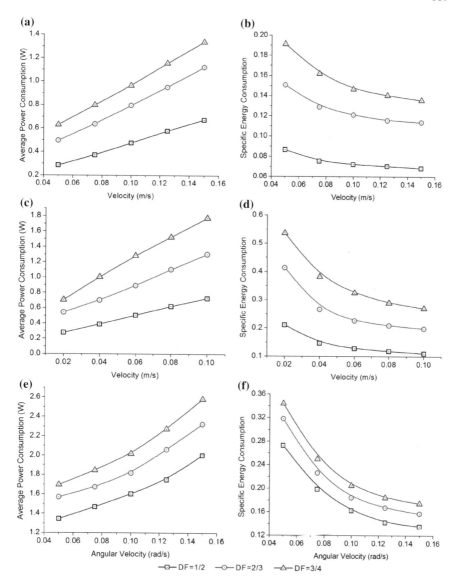

Fig. 4.16 Dependence of average power consumption (P_{av}) and specific power consumption (E_s) on velocity of trunk body **a** P_{av} for straight-forward motion, **b** E_s for straight-forward motion, **c** P_{av} for crab motion, **d** E_s for crab motion, **e** P_{av} for turning motion, **f** E_s for turning motion. Sloping angle (α_G) of terrain = 30° (considered for straight-forward motion); Banking angle (β_G) of terrain = 30° (considered for crab and turning motion)

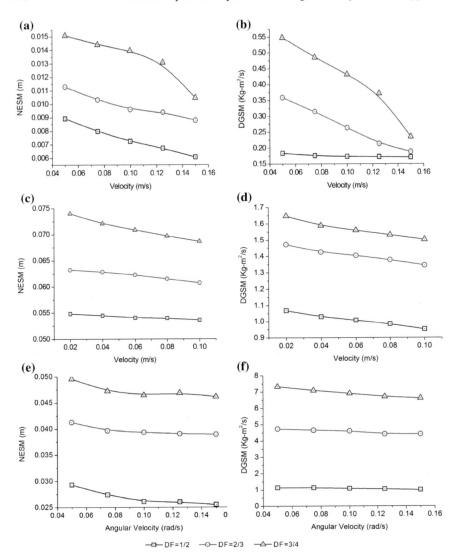

Fig. 4.17 Variations of net energy stability margin (NESM) and dynamic gait stability measure (DGSM) with velocity of trunk body **a** NESM for straight-forward motion, **b** DGSM for straight-forward motion, **c** NESM for crab motion, **d** DGSM for crab motion, **e** NESM for turning motion, **f** DGSM for turning motion. Sloping angle (α_G) of terrain = 30° (considered for straight-forward motion); Banking angle (β_G) of terrain = 30° (considered for crab and turning motion)

factor, thereby resulting in the higher angular velocity and angular acceleration of the system, and consequently, the energy consumption became more.

Another issue that has an effect on the power consumption with the increase in duty factor is the number of grounded legs. More number of grounded legs means the system has to overcome more frictional forces during its movement. The trend complies with the studies reported by Lin and Song [20]. Further, the curves plotted for different duty factors in Fig. 4.16b, d, and f shows that there is a maximum velocity of the trunk body for which E_s is a minimum. However, due to dynamic constraints of the joint actuators, the linear or angular velocity of the robot cannot be increased beyond a limit.

Figure 4.17 shows the plotted graphs related to the stability of the system (both NESM and DGSM) that varies with trunk body velocity for different duty factors. It is seen that with the increase in trunk body velocity of the system, the NESM decreases which means the system is subjected to instability with the increase in velocity. On the other hand, DGSM of the system is also seen to decrease as velocity of the system increases. This has occurred because the time period of each locomotion cycle decreases with the increase in trunk body velocity, thereby shortening the time period of the dynamic gaits. Now, due to decrease in the time period, the velocity of the swing leg increases which leads to higher angular momentum of the system at any instant, thereby, degrading the stability. The present study can be correlated with the studies reported by Lin and Song [20]; Garcia and Santos [11]. Also, it is found that for a particular trunk body velocity, NESM and DGSM increases with the increase in duty factor as expected.

4.2.2.2 Effect of Stroke on Energy Consumption and Stability

The computed results of the performance indices (here, energy consumption and stability) with respect to the body stroke for three locomotion cycles during straight-forward, crab, and turning motions of the robot with wave gaits of different duty factors are plotted in Figs. 4.18 and 4.19, respectively. The ranges of body strokes have been kept different for different type of motion, as shown in the Figs. 4.18 and 4.19, and had been chosen after closely studying the workspace of the legs. It is observed that increase in the body stroke increases both P_{av} and E_s, respectively, for all the three types of motion and duty factors for a specified input velocity, body height, and leg offset. This is due to the fact that the robot's leg during swing and support phase translates more, that is, covers more distance for larger strokes compared to that for shorter strokes. Again, with the increase in stroke, the angular velocity and angular acceleration of the joints also increase, which means higher joint torques. This ultimately leads to increase in energy consumption. Also, it is found that (refer to Fig. 4.18) for a particular body stroke, the energy consumption (both P_{av} and E_s) increases with the increase in duty factor. This is attributed to the fact that at a particular body stroke, the time period of each locomotion decreases with the increase in duty factor which means the angular velocity and angular acceleration of the joints are increased for movement of the legs in a synchronous manner. Besides

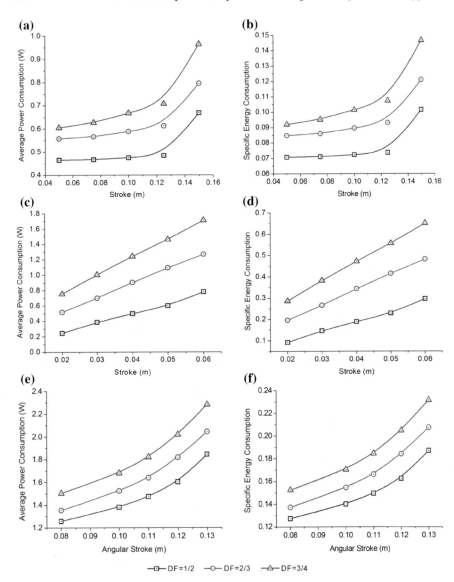

Fig. 4.18 Dependence of average power consumption (P_{av}) and specific power consumption (E_s) on body stroke **a** P_{av} for straight-forward motion, **b** E_s for straight-forward motion, **c** P_{av} for crab motion, **d** E_s for crab motion, **e** P_{av} for turning motion, **f** E_s for turning motion. Sloping angle (α_G) of terrain = 30° (considered for straight-forward motion); Banking angle (β_G) of terrain = 30° (considered for crab and turning motion)

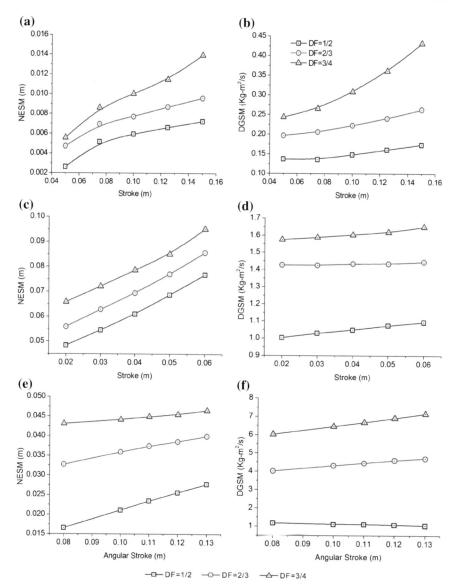

Fig. 4.19 Variations of net energy stability margin (NESM) and dynamic gait stability measure (DGSM) with body stroke **a** NESM for straight-forward motion, **b** DGSM for straight-forward motion, **c** NESM for crab motion, **d** DGSM for crab motion, **e** NESM for turning motion, **f** DGSM for turning motion. Sloping angle (α_G) of terrain = 30° (considered for straight-forward motion); Banking angle (β_G) of terrain = 30° (considered for crab and turning motion)

that, with the increase in duty factor, more number of legs are in contact with the ground. This means more frictional force is required to be overcome for the motion of the robot.

The effects of body stroke (linear/angular) on the minimum value of NESM and DGSM are shown in Fig. 4.19. It is seen that the minimum values of NESM increase with the increase in stroke which agrees with the work of McGhee and Frank [26] for static gait stability of the robot. Similar characteristics have been observed for the overall DGSM that corresponds to dynamic gait stability and are in agreement with the work carried out by Koo and Yoon [17]. Moreover, a commonly known fact that gait stability increases with increase in duty factor (for both static and dynamic gaits) had also been established through the graphs plotted in Fig. 4.19.

4.2.2.3   Effect of Body Height on Energy Consumption and Stability

The effects of variation of the trunk body height on energy consumption and stability have been carried out in the present study. The robot locomotes on various terrains, like up the hill during straight-forward motion ($\alpha_G = 30°$) and on banked surface ($\beta_G = 30°$) during crab and turning motion with wave gaits of different duty factors, as mentioned earlier. The variations of energy consumption (i.e., P_{av} and E_s) and stability (i.e., NESM and DGSM) with trunk body height are shown in Fig. 4.20 and 3.21, respectively. The graphs in Fig. 4.20 show that for a particular duty factor, as the height of trunk body increases, both the average power consumption and specific energy consumption of the system increase (Luneckas et al. [21]; Roy and Pratihar [32]. The reason is that as the height of the trunk body increases, the legs are raised higher keeping other parameters (trunk body velocity, leg offset, stroke, maximum height of swing leg, etc.) constant. This means the joints of the legs have to move faster leading to increase in the velocity and acceleration of the joints. Hence, the joints experience more torque, thereby, leading to more energy consumption by the robot. Again, it is found that as the duty factor increases, the energy consumption also increases for a particular height of the trunk body. This is due to the fact that for a particular body height, with increase in duty factor, the time period of each cycle decreases, leading to higher angular velocity and acceleration in the joints to synchronize the robot's motion. Also, it is seen that with the increase in duty factor, the number of grounded legs increases, which means more frictional forces in the leg tips are required to overcome the robot's motion, though the force distribution in the feet reduces.

Figure 4.21 shows the variations of minimum values of NESM and DGSM with the change in the height of the trunk body during straight-forward, crab, and turning motions with three different duty factors. It has been noticed that for a particular duty factor, if the parameters like trunk body velocity, leg offset distance, and trunk body stroke are kept constant, the minimum values of NESM and DGSM decrease with the increase in trunk body's height. This leads to instability in the system of the robot due to the increase in COG of the system in the support plane formed by the supporting legs. The study is found to conform with the work of Garcia et al. [12].

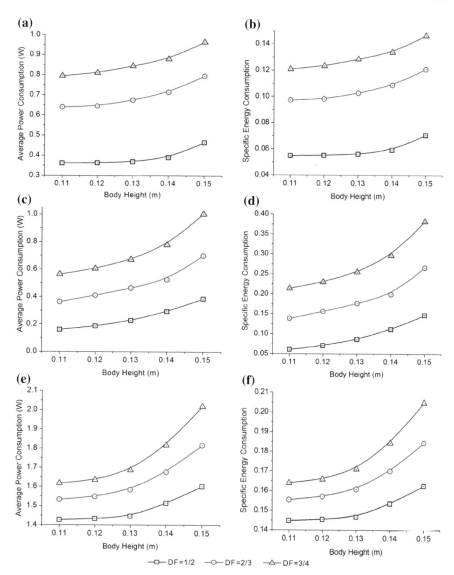

Fig. 4.20 Dependence of average power consumption (P_{av}) and specific power consumption (E_s) on the height of the trunk body **a** P_{av} for straight-forward motion, **b** E_s for straight-forward motion, **c** P_{av} for crab motion, **d** E_s for crab motion, **e** P_{av} for turning motion, **f** E_s for turning motion. Sloping angle (α_G) of terrain = 30° (considered for straight-forward motion); Banking angle (β_G) of terrain = 30° (considered for crab and turning motion)

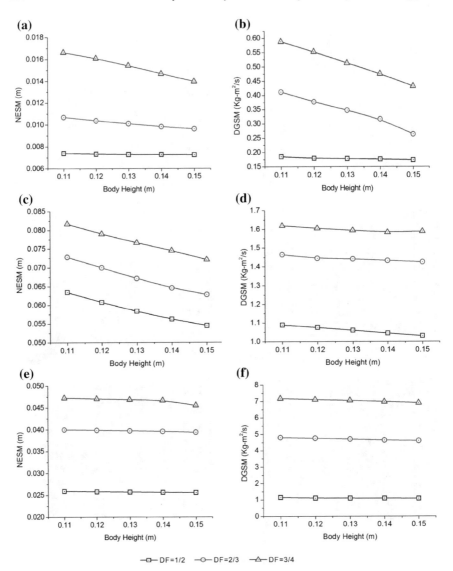

Fig. 4.21 Variations of net energy stability margin (NESM) and dynamic gait stability measure (DGSM) with height of the trunk body **a** NESM for straight-forward motion, **b** DGSM for straight-forward motion, **c** NESM for crab motion, **d** DGSM for crab motion, **e** NESM for turning motion, **f** DGSM for turning motion. Sloping angle (α_G) of terrain = 30° (considered for straight-forward motion); Banking angle (β_G) of terrain = 30° (considered for crab and turning motion)

4.2.2.4 Effect of Leg Offset on Energy Consumption and Stability

The effect of change in leg offset distance for energy consumption and stability has been carried out for three different types of motion and duty factors (DF) = 1/2, 2/3, and 3/4. As shown in Fig. 4.22, the values of P_{av} and E_s increase with the increase in lateral/radial offset distance for a particular duty factor (Santos et al. [33]; Lin and Song [20]. The reason behind is that, a finite number of reaction force vectors that cross the hip/knee joint are reduced with increase in offset distance. Further, to synchronize the motion of the robot, the magnitude of joint velocities and accelerations is found to increase with the increase in leg offset (all other parameters are kept constant).

Consequently, the joint torque values increase leading to more energy consumption. It is also to be noted that (refer to Fig. 4.22) for a fixed offset distance, the energy consumption increases with the increase in duty factor due to reduced value of time period of each cycle (all other parameters like body stroke, body height, and trunk body velocity are kept constant).

From the graphs plotted in Fig. 4.23, it is observed that the minimum values of NESM and DGSM increase as the offset distance increases for a particular duty factor. This is due to the increases in the projected foothold area of the support legs on a plane with the increase leg offset. This helps the COG of the system to move around in a more wider space.

4.2.2.5 Effect of Variable Geometry of Trunk Body on Energy Consumption and Stability

On the basis of geometrical configuration, six-legged robots are classified into two categories, namely (a) rectangular with bilateral symmetry (Erden and Leblebicioglu [8, 9]; Roy and Pratihar [31] and (b) hexagonal with radial symmetry (Wang et al. [40]. Energy consumption, stability of such types of robots with typical gait and path planning has been studied. In the present case, an attempt has been made to study the effects of geometrical variations of the trunk body on energy consumption and stability for straight-forward, crab, and turning motions. The geometry of the trunk body is varied by shifting the locations of joint 1 of leg 3 and leg 4 to the points S_3 and S_4, respectively, thereby changing the configuration of the robot from rectangular to hexagonal. It is varied by 'Δx' along X-direction of the local reference frame L_0, as shown in Fig. 4.24.

Computer simulations are run for three cycles in MATLAB with the necessary inputs as mentioned previously. The mass of the trunk body is kept the same for all values of shift (Δx) to study their effects on power consumption and stability of the system. It is observed from the graphs plotted in Fig. 4.25 that the average power consumption and specific power consumption decreased due to the said shifting. Study also showed that the maximum reduction in energy expenditure for the hexagonal body shape (i.e., at $\Delta x = 0.1$ m) is about 12.5%, 7.1%, and 6.1% with respect to rectangular body shape (i.e., at $\Delta x = 0.0$ m) for straight-forward motion, crab,

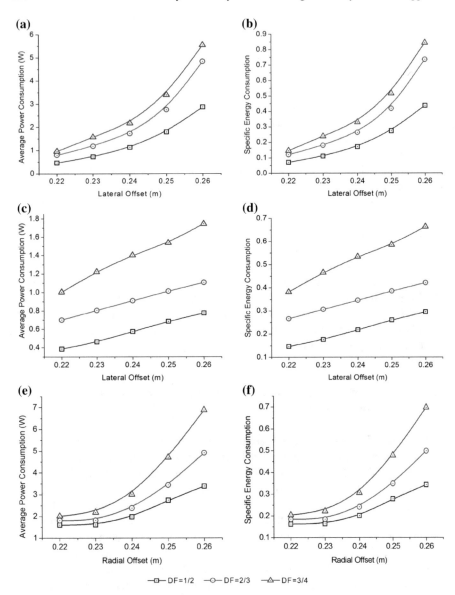

Fig. 4.22 Dependence of average power consumption (P_{av}) and specific power consumption (E_s) on leg offset **a** P_{av} for straight-forward motion, **b** E_s for straight-forward motion, **c** P_{av} for crab motion, **d** E_s for crab motion, **e** P_{av} for turning motion, **f** E_s for turning motion. Sloping angle (α_G) of terrain = 30° (considered for straight-forward motion); Banking angle (β_G) of terrain = 30° (considered for crab and turning motion)

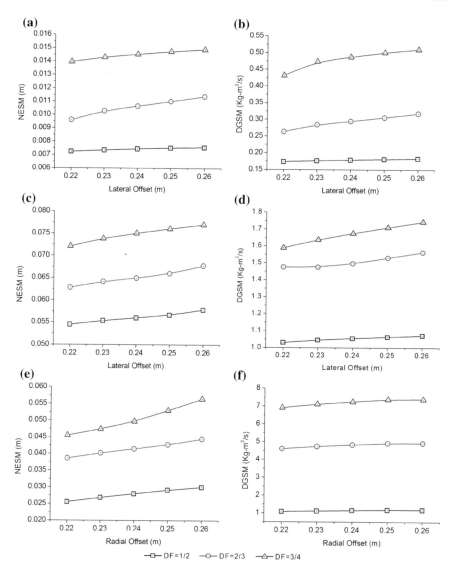

Fig. 4.23 Variations of net energy stability margin (NESM) and dynamic gait stability measure (DGSM) with leg offset **a** NESM for straight-forward motion, **b** DGSM for straight-forward motion, **c** NESM for crab motion, **d** DGSM for crab motion, **e** NESM for turning motion, **f** DGSM for turning Motion. Sloping angle (α_G) of terrain = 30° (considered for straight-forward motion); Banking angle (β_G) of terrain = 30° (considered for crab and turning motion)

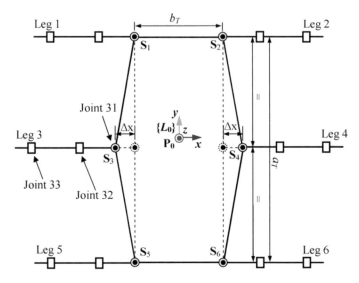

Fig. 4.24 Schematic top view of the six-legged robot showing shift of joint location S_3 and S_4 by Δx

and turning motion, respectively. The is due to the fact that as the shape of the body changes from rectangular to hexagonal, distribution of support-foot forces improved, thereby, leading to better torque distribution in the joints. Hence, the energy efficiency also improved. In this regard, the study of Santos et al. [33] is noteworthy, although the study is restricted to only a particular type of hexagonal configuration instead of varying it as carried out in the present study. Again, with the increase in duty factor (DF = ½–3/4), the power consumption is found to increase. This could be attributed to the fact that as DF increases, the time period of each cycle decreases for that particular Δx leading to increase in angular velocity and acceleration of the joints to synchronize the robot's motion. Moreover, with increase in duty factor, more number of legs comes to support phase at any instant of time, thereby, leading to more frictional forces that are required to overcome for the movement of the robot.

However, the stability margin of the system is found to increase with the increase in the value of Δx. It is observed in Fig. 4.26 that the minimum values of NESM and DGSM increase with the increase in Δx. The reason is that as Δx increases, the support area of the polygon formed by the legs in support phase increases. Also, the stability increases with the increase in duty factor, since more number of legs are in the support phase with the increase in the duty factor. The study can be correlated with the work carried out by Preumont et al. [30] and Takahashi et al. [38] which showed that stability margin of statically stable six-legged robots with hexagonal body shapes is better compared to rectangular robots.

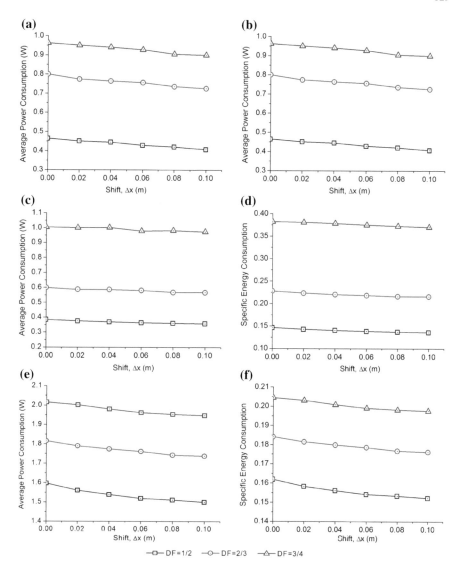

Fig. 4.25 Dependence of average power consumption (P_{av}) and specific power consumption (E_s) on variable geometry of trunk body **a** P_{av} for straight-forward motion, **b** E_s for straight-forward motion, **c** P_{av} for crab motion, **d** E_s for crab motion, **e** P_{av} for turning motion, **f** E_s for turning motion. Sloping angle (α_G) of terrain $= 30°$ (considered for straight-forward motion); Banking angle (β_G) of terrain $= 30°$ (considered for crab and turning motion)

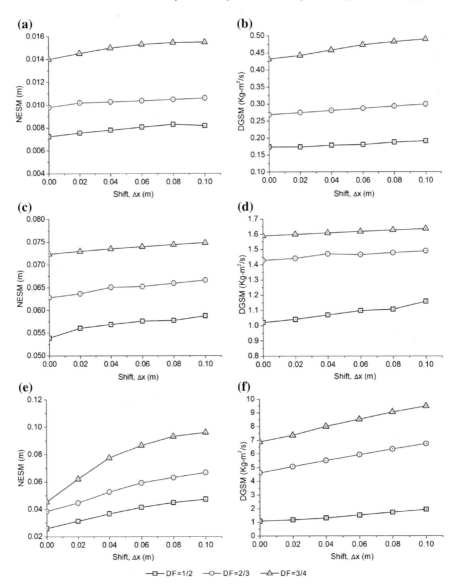

Fig. 4.26 Variations of net energy stability margin (NESM) and dynamic gait stability measure (DGSM) with variable geometry of trunk body **a** NESM for straight-forward motion, **b** DGSM for straight-forward motion, **c** NESM for crab motion, **d** DGSM for crab motion, **e** NESM for turning motion, **f** DGSM for turning motion. Sloping angle (α_G) of terrain = 30° (considered for straight-forward motion); Banking angle (β_G) of terrain = 30° (considered for crab and turning motion)

4.2.2.6 Effect of Crab Angle on Energy Consumption and Stability

In the present study of crab motion analysis, the effect of crab angle variation on energy consumption and stability has also been studied, in addition to the other variable parameters discussed in Sects. 4.2.2.1–4.2.2.5. The minimum crab angle is determined from the geometrical constraints of the robot and is limited to 20°.

It is observed (refer to Fig. 4.27) that at a particular duty factor, if the parameters like body stroke, body height, trunk body velocity, leg offset, etc., are kept constant, the average power consumption and specific energy consumption increase with the increase in crab angle from 20° to 90°. This is due to the fact that as the crab angle increases, the trunk body velocity component along Y increases for the same X-component of its velocity as per Eq. (3.66). Subsequently, the joint velocity and acceleration increase to synchronize the gait sequence. This leads to the increase in joint torque and energy consumption of the robot. Further, it is found that at lower crab angles, the rate of increase in P_{av} and E_s is less compared to that at higher crab angles, as seen from Fig. 4.27.

Further, the energy consumption is found to increase with the increase in duty factor. This is due to the fact that for a specific body stroke and velocity, if the duty factor increases, the time period decreases. This leads to increase in velocity and acceleration of the joints which in turn increases the energy consumption. The study can be correlated with the investigation done by Sanz-Merodio et al. [36]. It showed that the increase in crab angle increases the mechanical power consumption of six-legged robots.

Like energy consumption analysis, stability (both statically and dynamically stable) of the system is also carried out to calculate the changes in NESM and DGSM (refer to Fig. 4.28) with respect to crab angle. It is found that the robot is statically and dynamically stable for the wave gaits and environment under consideration. Further, it is seen that the stability (both NESM and DGSM) is seen to increase with

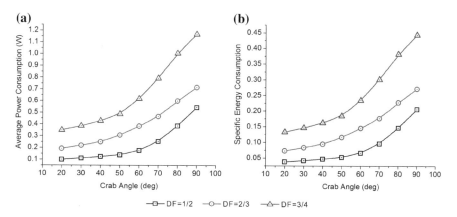

Fig. 4.27 Dependence of average power consumption (P_{av}) and specific power consumption (E_s) on crab angle **a** P_{av} **b** E_s. Banking angle (β_G) of terrain = 30°

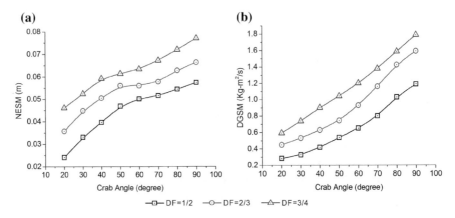

Fig. 4.28 Variation of net energy stability margin (NESM) and dynamic gait stability measure (DGSM) with crab angle **a** NESM **b** DGSM banking angle (β_G) of terrain $= 30°$

crab angle for a particular duty factor. It could be explained that as the crab angle increases, the projected foothold area distribution on either side of the motion axis of the robot increases during support phase. Also, it is seen that stability increases as duty factor increases, which is a known fact, since there are more number of legs in support phase at any instant of time during the robot's locomotion.

4.3 Summary

In this work, a generalized analytical coupled multi-body inverse dynamics model of a realistic six-legged robot locomoting on various terrains is developed. The multi-body dynamic modeling consists of rigid multi-body kinematic linkages, wherein the coupling effects have been considered along with the compliant contact impact and slip of the feet tip of support legs. The coupled dynamic model used in the present investigation can effectively tackle the inertia effects of swing leg on the support legs, which has not been addressed before. A large number of kinematic constraint equations. (114 equations) in terms of 3D Cartesian coordinates have been effectively reduced to 24 constraint equations in terms of generalized coordinates using kinematic transformation algorithms for efficient computation. Feasible solutions are obtained using quadratic optimization function for the contact dynamic model with consideration of impact and friction models. The quadratic optimization problem has been formulated as minimization of the sum of the squares of joint torques subjected to both equality and inequality constraints for efficient distribution of feet forces in the support legs, joint torques, energy consumptions etc. The performance of the developed model has been tested using computer simulations. It is observed that a momentarily impulsive force is generated when the foot tip collides with the terrain at the instant when the leg comes to support phase from swing phase. A sudden shoot in the normal reaction force is observed. The study of the analytical data and the graphs also show the effects of impact on the joint torques of its own leg and that of others due to coupling effects. Moreover, it is also observed that the middle revolute joint of each leg experiences more torque compared to the other joints at any instant of time. Other findings include the higher values of joint torques during support phase compared to that during swing phases for all the legs which is obvious, since the weight of the trunk body and payload are transferred to the ground through these support legs.

Simulation results for the study of energy consumption (*average power consumption* and *specific energy consumption*) and stability (*net energy stability margin* and *dynamic gait stability measure*) for the straight-forward, crab, and turning motions of the six-legged robot with three different duty factors, namely DF = 1/2, 2/3, and 3/4 on various terrain conditions have showed that a less power consumption and an increase in stability of the robot can be achieved by an appropriate selection of gait (i.e., duty factor) and its affecting parameters like walking velocity, body height, body stroke, leg offset, varying trunk body geometry from rectangular to hexagonal, and crab angle on various terrain conditions. Like, it is found that average power consumption increased with the increase in either of the walking parameters as mentioned above. However, the average power consumption decreased with the increase in transverse shift of the middle joints. On the other hand, stability (both NESM and DGSM) decreased with increase in body velocity or body height, but increased with increase in body stroke or leg offset or crab angle or transverse shift of the middle joints. Based on the plots, one could choose a proper duty factor and gait parameters to ensure the minimum energy consumption and positive gait stability.

References

1. J.P. Barreto, A.Trigo, P.Menezes, J. Dias, A.T. De Almeida, FBD-the Free body diagram method. kinematic and dynamic modeling of a six leg robot. in *Fifth International Workshop on Advanced Motion Control (AMC'98),* (Coimbra, Portugal, 1998) pp. 423–428
2. G. Bastos Jr., R. Seifried, O. Brüls, Inverse dynamics of serial and parallel underactuated multibody systems using a DAE optimal control approach. Multibody Sys. Dyn. **30**, 359 376 (2013)
3. M. Bennani, F. Giri, Dynamic modelling of a four-legged robot. J. Intell. Rob. Syst. **17**(4), 419–428 (1996)
4. G. Bessonnet, J. Marot, P. Seguin, P. Sardain, Parametric-Based dynamic synthesis of 3D-Gait. Robotica **28**(4), 563–581 (2010)
5. Q. Bombled, O. Verlinden, Dynamic simulation of six-legged robots with a focus on joint friction. Multibody Sys. Dyn. **28**(4), 395–417 (2006)
6. J.-S. Chen, F.-T. Cheng, K.-T. Yang, F.-C. Kung, S. York-Yih, Optimal force distribution in multilegged vehicles. Robotica **17**(2), 159–172 (1999)
7. F.T. Cheng, D.E. Orin, Optimal force distribution in multiple chain robotic systems. IEEE Trans. Syst. Man Cybern. **21**(1), 13–24 (1991)
8. M.S. Erden, K. Leblebicioglu, Torque distribution in a six-legged robot. IEEE Trans. Rob. **23**(1), 179–186 (2007)
9. M.S. Erden, K. Leblebicioglu, Analysis of wave gaits for energy efficiency. Auton. Robots **23**, 213–230 (2007)

10. E.F. Fukushima, S. Hirose, *Attitude and Steering Control of the Long Articulated Body Mobile Robot KORYU*, ed by H. Zhang. Climbing and Walking Robots: Towards New Applications (I-Tech Education and Publishing, 2007)
11. E. Garcia, P.G. Santos P. G. de, Controlling dynamic stability and active compliance to improve quadrupedal walking. in *Proceedings of the 8th International Conference on Climbing and Walking Robots and the Support Technologies for Mobile Machines (CLAWAR 2005)*, (2008) pp. 205–212
12. E. Garcia, J. Estremera, P.G. de Santos, A comparative study of stability margins for walking machines. Robotica **20**(6), 595–606 (2002)
13. H. Hahn, *Rigid Body Dynamics of Mechanisms 1*, 1st edn. (Springer-Verlag, Berlin-Heidelberg, 2002)
14. S. Hirose, H. Tsukagoshi, K. Yoneda K, Normalized energy stability margin: generalized stability criterion for walking vehicles. in *Proceedings of the International Conference on Climbing and Walking Robots* (Brussels, Belgium, 1998) pp. 71–76
15. K. Hunt, F. Crossley, Coefficient of restitution interpreted as damping in vibroimpact. Trans ASME-J. Appl. Mech. **42**, 440–445 (1975)
16. D.C. Kar, K.K. Issac, K. Jayarajan, Minimum energy force distribution for a walking robot. J. Robotic Syst. **18**(2), 47–54 (2001)
17. T.-W. Koo, Y.-S. Yoon, Dynamic instant gait stability measure for quadruped walking robot. Robotica **17**, 59–70 (1999)
18. Z. Li, S.S. Ge, S. Liu, S, Contact-force distribution optimization and control for quadruped robots using both gradient and adaptive neural networks. IEEE Trans. Neural Netw. Learn. Syst. **25**(8), 1460–1473 (2014)
19. J. Liang, L. Xin, Dynamic Simulation of Spiral Bevel Gear based on SolidWorks and ADAMS. J. Theor. Appl. Information Technol. **47**(2), 755–759 (2013)
20. B.-S. Lin, S.-M. Song, Dynamic modeling, stability and energy efficiency of a quadrupedal walking machine. J. Robotic Syst. **18**(11), 657–670 (2001)
21. M. Luneckas, T. Luneckas, D. Udris, N.M.F. Ferreira, Hexapod robot energy consumption dependence on body elevation and step height. Elektron. IR Elektrotechnika **20**(7), 7–10 (2014)
22. M. Machado, P. Moreira, P. Flores, H.M. Lankarani, Compliant contact force models in multi-body dynamics: Evolution of the Hertz contact theory. Mech. Mach. Theory **53**, 99–121 (2012)
23. A. Mahapatra, S.S. Roy, D.K. Pratihar, Computer aided modeling and analysis of turning motion of hexapod robot on varying terrains. Int. J. Mech. Mater. Design **11**(3), 309–336 (2015)
24. A. Mahapatra, S.S. Roy, D.K. Pratihar, Study on feet forces' distributions, energy consumption and dynamic stability measure of hexapod robot during crab walking. Appl. Math. Model. **65**, 717–744 (2019)
25. D.W. Marhefka, D.E. Orin, Gait planning for energy efficiency in walking machines. in *Proceedings of IEEE International Conference on Robotics and Automation,* (Albuquerque, New Mexico, 1997) pp. 474–480
26. R.B. McGhee, A.A. Frank, On the stability properties of quadruped creeping gaits. Math. Biosci. **3**, 331–351 (1968)
27. D. Messuri, C. Klein, Automatic body regulation for maintaining stability of a legged vehicle during rough-terrain locomotion, IEEE J. Robot. Autom. RA-**1**(3), 132–141 (1985)
28. MSC.ADAMS Documentation and Help User Guide, https://simcompanion.mscsoftware.com/infocenter. MSC Software Corporation
29. E. Otten, Inverse and forward dynamics: models of multi-body systems. Phil. Trans. B, Royal Soc. London **358**(1437), 1493–1500 (2003)
30. A. Preumont, P. Alexadre, D. Ghuys, Gait analysis and implementation of a six legwalking machine. in 91ICAR, *Fifth International Conference on Advanced Robotics - Robots in Unstructured Environments* (Pisa, Italy, 1991) pp. 941–945
31. S.S. Roy, D.K. Pratihar, Dynamic modeling, stability and energy consumption analysis of a realistic six-legged walking robot. Rob. Comput. Integrated Manuf. **29**(2), 400–416 (2013)

32. S.S. Roy, D.K. Pratihar, Kinematics, dynamics and power consumption analyses for turning motion of a six-legged robot. J. Intell. Rob. Syst. **74**(3–4), 663–668 (2014)
33. P.G. de Santos, E. Garcia, R. Ponticelli, M. Armada, Minimizing energy consumption in hexapod robots. Adv. Robot. **23**(6), 681–704 (2009)
34. P.G. de Santos, J. Estremera, E. Garcia, Optimizing leg distribution around the body in walking robots. in *Proceedings of IEEE International Conference on Robotics and Automation* (Barcelona, Spain, 2005), pp. 3207–3212
35. P.G. de Santos, E. Garcia, J. Estremera, *Quadrupedal Locomotion-An Introduction to the Control of Four-legged Robots* (Springer-Verlag Limited, London, 2006)
36. D. Sanz-Merodio, E. Garcia, P.G. de Santos, Analyzing energy-efficient configurations in hexapod robots for demining applications. Ind. Robot Int. J. **39**(4), 357–364 (2012)
37. W.J. Stronge, *Impact Mechanics* (Cambridge University Press, 2000)
38. Y. Takahashi, T. Arai, Y. Mae, K. Inoue, N. Koyachi, Development of multi-limb robot with omnidirectional manipulability and mobility. in *Proceedings of the IEEE RSJ International Conference on intelligent Robots and Systems* (Takamatsu, Japan, 2000), pp. 2012–2017
39. H. Takemura, M. Deguchib, J. Uedaa, Y. Matsumotoa, T. Ogasawaraa, Slip-adaptive walk of quadruped robot. Robot. Auton. Syst. **53**(2), 124–141 (2005)
40. Z.Y. Wang, X.L. Ding, A. Rovetta, A. Giusti, Mobility analysis of the typical gait of a radial symmetrical six-legged robot. Mechatronics **21**(7), 1133–1146 (2011)

Chapter 5
Validation Using Virtual Prototyping Tools and Experiments

In this chapter, a few case studies are carried out to validate the analytical results based on developed formulation in the book with the simulated results in MSC.ADAMS® (Automated Dynamic Analysis of Mechanical Systems), which is a commercially available rigid multi-body dynamics numerical solver. The previously developed models are not validated, and if done, it is totally experimental, which is time-consuming and might not be cost effective. The chapter provides implementation of the developed algorithms on a virtual model of a six-legged robot and improving its performance using VP technology prior to developing its physical prototype. The present model of the realistic six-legged robot has emphasized upon all such issues for negotiating various terrains. Further, validation using experiments has also been carried out to prove the robustness of the developed algorithm.

5.1 Modeling Using Virtual Prototyping Tools

Today, the development of a product, right from the conceptual design stage has taken leapfrog worldwide due to tools like virtual reality (VR) and virtual prototyping (VP). These tools help designers and engineers to understand the full implications of product's technical requirements elaborated in the design and the associated analysis. Ullman [1] and Pratt [2] pointed out that up to 70% of the total life cycle costs of a product are committed by decisions made in the early design stage. In VP, CAD softwares are used to build the virtual prototypes, which are digital representation of the geometry and attributes of the products. Further, to replace or supplement physical testing of the prototypes, high-fidelity physics-based software tools, that is, computer aided engineering (CAE) softwares are used to test or analyze the performance of the product. This has proved to be much better way to design innovative products. The designers get immense opportunities to optimize their design, analyze them to find design flaws, and make changes well before a real prototype is made. The VP paradigm can be applied at all stages of a product development, from

© Springer Nature Singapore Pte Ltd. 2020
A. Mahapatra et al., *Multi-body Dynamic Modeling of Multi-legged Robots*,
Cognitive Intelligence and Robotics, https://doi.org/10.1007/978-981-15-2953-5_5

requirements definition to sustainment. Therefore, it can be inferred that, virtual prototyping enables shorter product development cycles, reduced physical prototyping
costs, better decision-making, and better quality of the products. This has been possible with the advancement in faster computation facilities where large problems can
be handled and well validated with high confidence level. The use of virtual prototyping by industries is rapidly growing. This offers them the potential for keeping
its edge over other organizations that are relying on the traditional "*design, build,
physical test, fix*" paradigm. The present work has taken the advantage of such tools
to develop and analyze its performance prior to the development of the system.

5.2 Numerical Simulation and Validation Using VP Tools and Experiments

The VP involves the following important steps: (1) development of the CAD model
of the hexapod using solid modeler CATIA V5, (2) defining the joints and contacts in
CATIA SimDesigner Workbench, (3) exporting as a MSC.ADAMS® compatible file
format, (4) pre-processing, execution, and post-processing of the results of analysis
in MSC.ADAMS® (refer to Mahapatra et al. [3]).

5.2.1 Validation of Kinematic Motion Parameters

The kinematic motion parameters obtained through computer simulations are
validated using VP tools.

5.2.1.1 Case Study 1: Crab Motion of the Robot to Avoid Obstacle on a Flat Terrain

The case study shows the typical application of crab motion during the robot's locomotion on a flat terrain with duty factor of 2/3. The robot moves in transverse direction
with a positive crab angle of 70° (refer to Fig. 5.1). Here, the X and Y-translational
velocity components of the trunk body are not independent unlike straight-forward
motion. Therefore, the translational velocity components of the trunk body with
respect to frame G are assumed to follow the relationship:

$$\dot{\mathbf{r}}_{P_0 O}^G = \left(\dot{x}_{P_0 O}^G, \dot{x}_{P_0 O}^G \cdot \cot \theta_c, 0\right)^{\mathrm{T}} \tag{4.1}$$

where ϑ_c is the crab angle.

At time $t = 0$, the position and orientation of $\mathbf{P}_0$ with respect to frame G are given
by $\mathbf{p}_0^G = \{0, 0.45, 0.15, -2, 2, 0\}^{\mathrm{T}}$. Assuming all the initial velocity components as

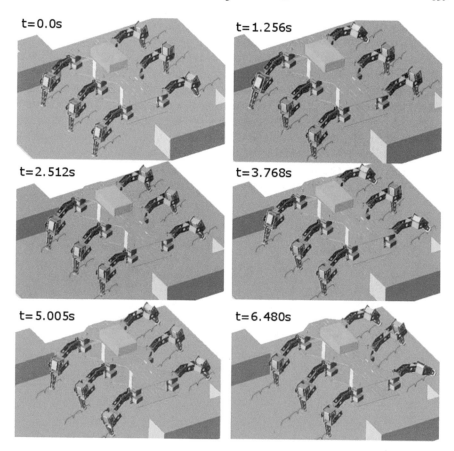

Fig. 5.1 Snapshots of a realistic six-legged robot simulated in ADAMS to avoid an obstacle on the way with crab motion capabilities using equal phase gait (DF = 2/3)

zero, the maximum translational and angular velocities reached by the trunk body are assumed to be as follows: $\dot{x}^G_{P_0O} = 0.02$ m/s, $\dot{\alpha}_0 = +0.01$ rad/s, $\dot{\beta}_0 = +0.02$ rad/s. In addition to the above, the other necessary inputs are $\boldsymbol{\eta}_G = (0, 0, 0)^T$, $\vartheta_{i1} = \pm(90° - \vartheta_c)$, $\beta_{i2} = \pm16°$, $\beta_{i3} = \pm69°$ (for $i = 1–6$), maximum height of swing along Z with respect to $\boldsymbol{G} = 0.015$ m. Also, the maximum slip velocity is assumed to be equal to 0.001 m/s with a slip angle of $\varepsilon_s = 45°$. The simulations are run for three duty cycles, i.e., $n = 3$ with DF = 2/3 and $s_0 = 0.02$ m, in MATLAB. Once calculated in MATLAB, the relevant motion data of the trunk body and leg tip are imported into MSC.ADAMS® as the inputs.

The simulations are run in MSC.ADAMS® solver for a total cycle time of 6.5 s with a time step of $h = 0.02$ s, as obtained using Eq. (2.92). Figure 5.1 displays some snapshots (interactive display frames) of the robot (simulated in MSC.ADAMS®), while walking on a flat terrain with crab motion capabilities to avoid an obstacle (equal phase gait strategy with DF = 2/3). The analytical solutions for various kinematic parameters (both translational and angular displacements, velocities, accelerations, etc.) as obtained by computing in MATLAB are validated with the simulated results of MSC.ADAMS®. Both the results are in close proximity of one another, which validates the mathematical model developed for crab motion. Figure 5.2 shows the comparative graphs of angular parameters (displacement, velocities, and accelerations) for various joints of leg 1. It is interesting to note that angular displacement of joint 11 is not kept constant (as expected for a constant crab angle θ_c) due to the presence of angular motions in the trunk body along X and Y axes with respect to frame L_0 (refer to Fig. 5.2a). Moreover, the other joint angles are found to be within the expected limits. The differences in the values of angular displacement of the legs vary in the range of $(-0.4°, 4°)$ approximately.

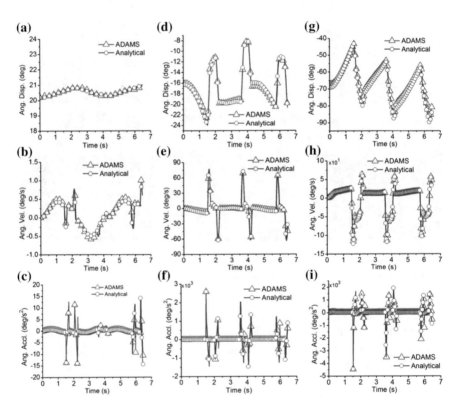

Fig. 5.2 Comparative graphs of the kinematic analysis of a realistic six-legged robot during crab motion using equal phase gait (DF = 2/3). Joint 11 **a** Angular displacement, **b** Angular velocity, **c** Angular acceleration. Joint 12, **d** Angular displacement, **e** Angular velocity, **f** Angular acceleration. Joint 13 **g** Angular displacement, **h** Angular velocity, **i** Angular acceleration

5.2.1.2 Case Study 2: Turning Motion of the Robot on a Banked Surface

In the present case study, an attempt has been made to maneuver the robot without payload over a banking surface using its turning motion capabilities with DF $= 1/2$. The rate of change of translational displacement (along X, Y, and Z) of the trunk body with respect to frame G follows the relationship, as given in Eq. (2.54).

At time $t = 0$, the position and orientation of P_0 with respect to frame G are given by $\mathbf{p}_0^G = \{1.3, 0.75, 0.15, 0, 0, 30\}^T$. The initial velocity components are assumed to be equal to zero. The maximum rate of change of angular displacement of the trunk body along Z direction with respect to frame L_0 is assumed to be equal to $\dot{\theta}_0\big|_{max} = \dot{\theta}_0\big|_{t=t_1} = +0.1$ rad/s with $\dot{\alpha}_0 = 0$, $\dot{\beta}_0 = 0$, in the present case. The turning radius of the trunk body throughout the motion of the robot is considered as $\rho_0 = 1.5$ m. The initial joint angles θ_{i1} of all the joints of the robot are calculated using some geometrical relations described in Appendix A.4. Here, all the joint angles are noted in the order of $[\theta_{i1}, \beta_{i2}, \beta_{i3}]$ (for $i = 1$–6) as $[5°, -16°, -69°]$ for leg 1, $[12°, 16°, 69°]$ for leg 2, $[12°, -16°, -69°]$ for leg 3, $[18.5°, 16°, 69°]$ for leg 4, $[22°, -16°, -69°]$ for leg 5, $[27.5°, 16°, 69°]$ for leg 6. In addition to the above, the other necessary inputs are $\boldsymbol{\eta}_G = [0, -15°, 0]^T$, maximum height of swing along Z with respect to frame $L_{i3} = 0.005$ m($= Hm_{min} + \Delta h$) for all duty cycles. Since the terrain surface is flat, $h'_{in} = 0$ for all duty cycles. Also, the maximum slip velocity is assumed to be equal to 0.001 m/s with a slip angle of $\varepsilon_s = 45°$. The simulations are run for three duty cycles, i.e., $n = 3$ with DF $= 1/2$ and $s_0^c = 0.120$ m in MATLAB.

The total time of motion of the robot for three duty cycles is calculated to be equal to 5.75 s (first cycle—2.05 s, second cycle—1.65 s, and the third cycle—2.05 s, respectively) by Eq. (2.92). The second cycle time is found to be less compared to the first and third cycle time due to the effects of acceleration and deceleration on the trunk body in the starting and ending of cycles, respectively. The variations of the angular displacement rates of the trunk body during the robot's motion along the banked surface are computed using Eqs. (2.41) and (2.51) followed by the computation of the translational displacement rates of the trunk body, based on the Eq. (2.54) (refer to Fig. 5.3a, b). It could be observed from Fig. 5.3a, b that the translational displacement rate along Z-axis and angular displacement rate along X and Y axes are zero. This means that the trunk body is always kept at a constant height with respect to the frame G in the present case study.

The position, velocity, and acceleration of the leg tip P_{i3} (i $= 1$–6) are computed with respect to frame G_0 based on the motion and gait planning algorithms, as discussed in Sects. 3.2.5 and 3.2.7. The trace of the position of P_{i3} (i $= 1$–6) is plotted in 3D Cartesian space, as shown in Fig. 5.4. The projections of the 3D motion trajectories of all the legs on the XY plane show the curve paths followed by the robot's legs to execute turning motion. Here, the effect of slip is negligible, since the slip velocity (in XY plane) is very less compared to the swing velocity of the legs. Figure 5.5 shows the translational velocity of the tip of leg 1 in XY plane of frame G during the support phase for a given gait sequence that illustrates the slippage. The maximum slip distance obtained from calculations is approximately equal to 0.001 m.

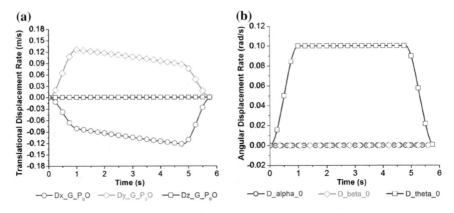

Fig. 5.3 Trunk body motion on a banked surface for three duty cycles **a** rate of change of translational displacement, **b** rate of change of angular displacement

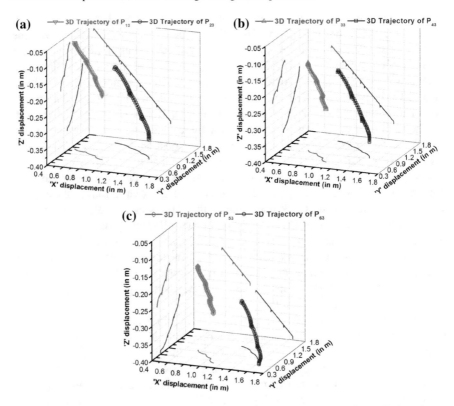

Fig. 5.4 3D motion trajectory of tip $\mathbf{P}_{i3}$ on a banked surface with respect to frame G_0 for **a** legs 1 and 2, **b** legs 3 and 4, **c** legs 5 and 6

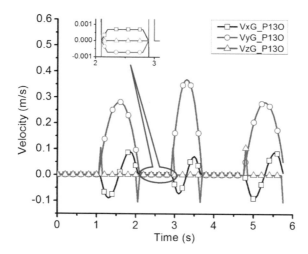

Fig. 5.5 Translational velocities of tip point P_{13} (leg 1) with respect to frame G, showing slip motion during support phase in *XY* plane

After the computations in MATLAB are over, the relevant motion data (velocity) of the trunk body and leg tip are imported into MSC.ADAMS®, as the inputs. The simulations are subsequently run in MSC.ADAMS® solver for a total cycle of 5.75 s with a time step of $h = 0.05$ s. Some of the snapshots of the robot simulated in MSC.ADAMS®, while traversing the banked surface using wave gait strategy and turning motion capabilities are shown in Fig. 5.6. The robot follows the gait sequences: leg 1–4–5 in swing phase and 2–3–6 in support phase during a half cycle. During the next half cycle, leg 1–4–5 is in support phase and 2–3–6 is in swing phase. The simulated results are in agreement with the computed data of the proposed motion planning algorithm, which further can prove its efficacy. The corresponding simulation time is indicated for each snapshot. The results of kinematic analysis of leg 1 obtained in MATLAB are compared with those of MSC.ADAMS®, and a close match is obtained (refer to Fig. 5.7). The joint angles are lying within the expected limits, which showed that at no time during the motion, the configuration of the robot is staggered and there is no interference among the legs during the motion. The differences in the values of angular displacement of the legs varied in the range of $(-3.8°, 9.1°)$ approximately. Also, the magnitude of angular velocities of joint 12 and 13 (refer to Fig. 5.7e, h) is less compared to that of joint 11 (refer to Fig. 5.7b). Moreover, Fig. 5.7b shows that the angular velocity of joint 11 during swing phase varies steadily at a faster rate compared to that during support phase. Maximum angular velocity of the joint occurred during swing phase of the cycle, which is due to the effect of trunk body motion on the swing motion of the robot's leg. Further, the position of aggregate COM of the system in 3D Cartesian space obtained analytically has been compared with the simulated results in MSC.ADAMS®. The analytical data are in close agreement with those obtained with MSC.ADAMS® (refer to Fig. 5.8).

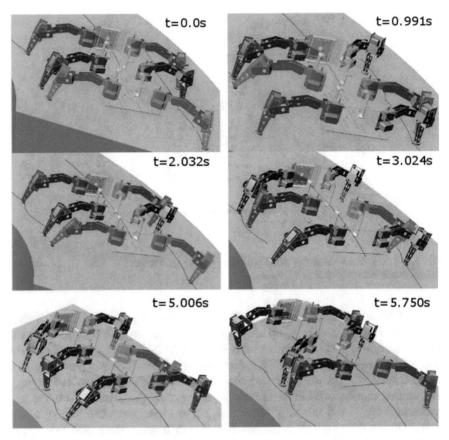

Fig. 5.6 Snapshots of a realistic six-legged robot simulated in MSC.ADAMS® for maneuverability over a banking surface with turning motion capabilities using wave gait (DF = 1/2)

5.2.1.3   Case Study 3: Turning Motion of the Robot in an Uneven Terrain

Locomotion analysis of a six-legged robot over an uneven terrain is complex. An attempt has been made in the present case study to explore the turning motion capabilities of the robot with payload on such terrains with DF = 1/2. The surface irregularities of the terrain are assumed to be small with values ranging from 5 to 8 mm with respect to local frame L_{i3} on the path of the leg swing. To make the motion of robot realistic on such topography, both translational and angular displacement rates of the trunk body are taken into consideration. The angular displacement rates of the trunk body are governed by functions $g_x(t)$, $g_y(t)$, and $g_z(t)$ along their respective axes, whereas the translational displacement rates (along X, Y, and Z) of the trunk body are governed by Eq. (3.54), as discussed in Sect. 3.2.5.1.

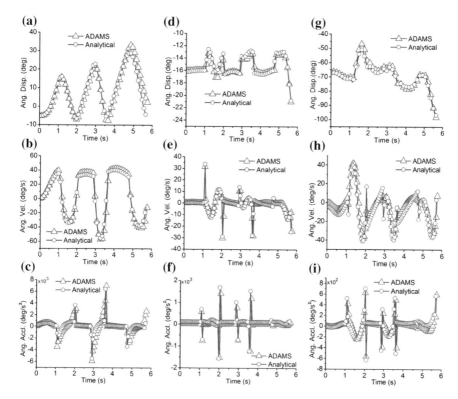

Fig. 5.7 Comparative graphs of the kinematic analysis of a realistic six-legged robot during turning motion on a banked surface using wave gait (DF = 1/2) for leg 1. Joint 11 **a** Angular displacement, **b** Angular velocity, **c** Angular acceleration. Joint 12 **d** Angular displacement, **e** Angular velocity, **f** Angular acceleration. Joint 13 **g** Angular displacement, **h** Angular velocity, **i** Angular acceleration

In the present case, at time $t = 0$ s, the position and orientation of the point $\mathbf{P}_0$ with respect to frame $\boldsymbol{G}$ are given by $\mathbf{p}_0^G = \{1.73, 1.0, 0.15, 1, -2, 30\}^T$. The initial velocity components of the bodies in the system are assumed to be equal to zero. The maximum and minimum rate of change of angular displacements of the trunk body along the axes X and Y with respect to frame $\boldsymbol{L}_0$ are given by $\dot{\alpha}_0|_{\max,\min} = \pm 0.01$ rad/s, $\dot{\beta}_0|_{\max,\min} = \pm 0.01$ rad/s, whereas the maximum rate of change of angular displacement of the trunk body along Z is assumed to be equal to $\dot{\theta}_0|_{\max} = \dot{\theta}_0|_{t=r_1} = +0.1$ rad/s. Similarly, the maximum rate of change of translational displacements of the trunk body along the respective axes with respect to frame $\boldsymbol{G}$ are given by $\dot{x}_{P_0O}^G|_{\max} = -\rho_0\dot{\theta}_0|_{\max}\sin\theta_0|_{t=t_1}$, $\dot{y}_{P_0O}^\sigma|_{\max} = -\rho_0\dot{\theta}_0|_{\max}\cos\theta_0|_{t=t_1}$ and $\dot{z}_{P_0O}^G|_{\max} = 0.005$ m/s (refer to Fig. 5.9). The turning radius (ρ_0) of the point $\mathbf{P}_0$ on the trunk body is assumed to be equal to 2.0 m. Also, the joint angle θ_{i1} ($i = 1$–6) for the robot's initial configuration is calculated using geometrical relations given in Appendix A.4. If the joint angles of leg i are in the order of $[\theta_{i1}, \beta_{i2}, \beta_{i3}]$ (for $i = 1$–6), then the corresponding joint angles are $[11.5°, -16°, -69°]$ for leg 1,

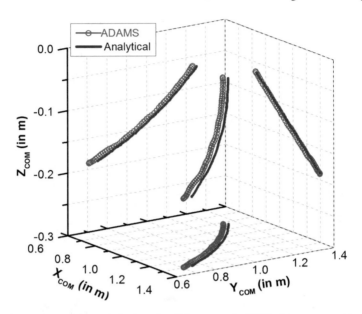

Fig. 5.8 Comparison of the analytical results of the aggregate COM of the system with respect to frame G_0 during turning motion on a banked surface. with that of MSC.ADAMS®

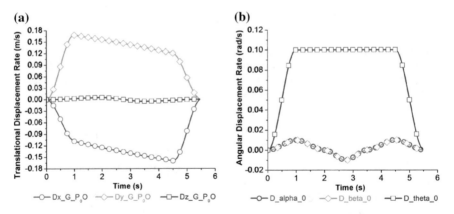

Fig. 5.9 Trunk body motion on an uneven terrain for three duty cycles **a** rate of change of translational displacement, **b** rate of change of angular displacement

[18°, 16°, 69°] for leg 2, [16.5°, −16°, −69°] for leg 3, [22.5°, 16°, 69°] for leg 4, [24°, −16°, −69°] for leg 5, [29.5°, 16°, 69°] for leg 6. The other relevant inputs are related to the topography of the terrain, namely (i) Hm_{in} and (ii) h'_{in} with respect to frame L_{i3} (refer to Fig. 3.4). For three duty cycles ($n = 3$), the values of Hm_{in} are in the order of $[Hm_{i1}, Hm_{i2}, Hm_{i3}]$ for $i = 1$–6, such that, [0.005, 0.005, 0.005] corresponds to leg 1, [0.006, 0.008, 0.003] corresponds to leg 2, [0.004, 0.005, 0.002] corresponds to leg 3, [0.006, 0.006, 0.004] corresponds to leg 4, [0.006, 0.008, 0.003] corresponds to leg 5, and [0.005, 0.005, 0.005] corresponds to leg 6 (all values are in m). Similarly, the values of h'_{in} are in the order of $\left[h'_{i1}, h'_{i2}, h'_{i3}\right]$ for $i = 1$–6, such that, [0.003, 0.005, -0.003] corresponds to leg 1, [0.004, 0.006, 0.0] corresponds to leg 2, [0.0042, 0.004, -0.002] corresponds to leg 3, [0.002, 0.002, -0.003] corresponds to leg 4, [0.0064, 0.006, 0.0] corresponds to leg 5, and [0.003, 0.005, 0.003] corresponds to leg 6 (all values are in m). The value of Δh is kept equal to 0.002 m. In addition to the above, the other necessary inputs are $\mathbf{\eta}_G = [0, 0, 0]^T$. Since, there is slippage of the legs during support phase, the maximum slip velocity is assumed to be equal to 0.005 m/s with a slip angle of $\varepsilon_s = 45°$ for all the legs. The simulations are run for three duty cycles with DF = 1/2 and $s_0^c = 0.150$ m in MATLAB.

The total simulation time computed in MATLAB to execute the motion of the robot for three duty cycles is 5.45 s (first cycle—2.0 s, second cycle—1.45 s, third cycle—2.0 s), as calculated using Eq. (2.92). It is to be noted that the robot completes the second cycle faster than the other two. This is due to the effects of acceleration and deceleration time of the robot during the first and third cycles, respectively. In the present case study, the vertical height of the trunk body is not kept constant and it varied with time due to the rate of change of translational displacement along Z and that of angular displacement in YZ and XZ planes (refer to Fig. 5.9).

The kinematic motion parameters (like position, velocity, and acceleration) of the tip point $\mathbf{P}_{i3}$ ($i = 1$–6) are calculated with respect to frame G_0 based on the motion and gait algorithm (refer to Sects. 3.2.5 and 3.2.7). The data points of the path followed by the robot's leg tip during turning motion are plotted in 3D Cartesian space, as shown in Fig. 5.10. Also, the projected data points on the XY plane showed the curved path followed by the robot during turning. The effect of slip velocity on the path of trajectory of the robot's leg is small, though there is a slippage of approximately 0.005 m in the XY plane during the support phase of the legs.

To validate the computed results in MATLAB, the velocity data of the trunk body and the leg tip $\mathbf{P}_{i3}$ are imported into MSC.ADAMS® as relevant inputs and preprocessed. The simulations are run in MSC.ADAMS® solver for the total cycle time of 5.45 s with a time step of $h = 0.05$ s, as computed in MATLAB. Visualization of the simulated data (refer to Fig. 5.11) of the robot maneuvering on an uneven terrain shows that the motion of the robot is stable and according to the desired sequence of 1–4–5 and 2–3–6 walking gait. This also proves the efficacy of the proposed motion planning algorithms. The corresponding simulation time is indicated for each snapshot. Moreover, the comparative study of the kinematic motion parameters of leg 6 shows that the results are in close agreement (refer to Fig. 5.12). The joint angles are within the expected limit, which also proves that the robot's configuration is not staggered and collision between the legs can be avoided during the robot's motion. The deviation of the angular displacement of leg 6 is in the approximate range of (5.9°, $-12.2°$). Another observation can be made from the plotted graphs of the angular velocities of joints 61, 62, and 63 (refer to Fig. 5.12b, e, h) like the magnitude of angular velocity of joints 62 and 63 is less compared to that of joint 61. A close look on Fig. 5.12b shows that during the swing phase, the angular velocity of joint 61 varies steadily at a faster rate compared to angular velocity of that joint

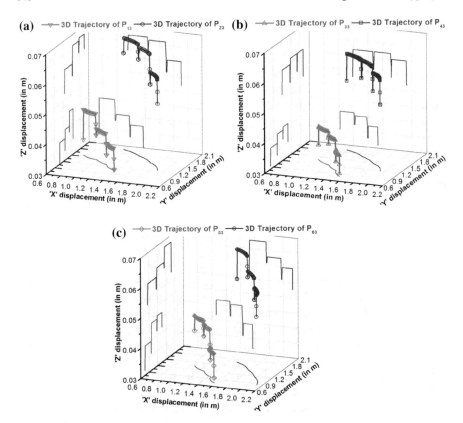

Fig. 5.10   3D motion trajectory of leg tip $\mathbf{P}_{i3}$ on an uneven terrain with respect to frame G_0 **a** legs 1 and 2, **b** legs 3 and 4, **c** legs 5 and 6

during support phase. Moreover, the maximum angular velocity of the joint occurs during the swing phase of a cycle. This is due to the effect of trunk body motion on the swing motion of the robot's leg. So, the joint $i1$ ($i = 1$–6) is the most predominate joint, since it controlled the motion sequence of the legs of the robot.

Further, comparative analysis of the displacement of the aggregate COM of the robot with payload is carried out. The results are plotted in Fig. 5.13 to show that the analytical results are in close agreement with the MSC.ADAMS® results. Moreover, the COM varied in 3D Cartesian along X, Y, and Z with respect to frame G_0, which gives a realistic picture of the robot's motion in various terrains. The results further prove the efficacy of the method of analysis introduced in this contribution for the kinematics of the system.

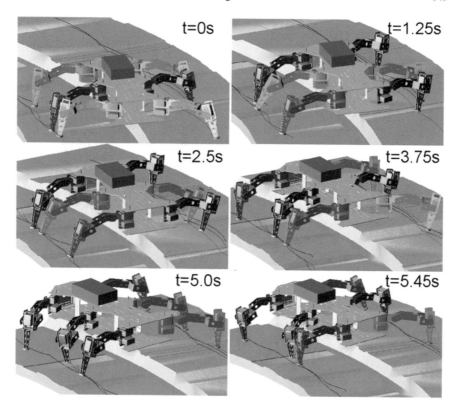

Fig. 5.11 Snapshots of a realistic six-legged robot simulated in MSC.ADAMS® for maneuverability over an uneven terrain with turning motion capabilities using wave gait (DF = 1/2)

5.2.2 Validation of Dynamic Motion Parameters

The dynamic motion parameters obtained through computer simulations are validated using VP tools and experiments.

5.2.2.1 Case Study 1: Staircase Climbing of the Robot with Straight-Forward Motion

In the present case study, maneuverability of the realistic six-legged robot climbing up a staircase (breadth, $b_s = 0.25$ m; height, $h_s = 0.05$ m) has been analyzed with DF = 1/2. The kinematic analysis of the system has been performed with predefined parameter inputs using MATLAB solver and thereafter, multi-body dynamic analysis of the system is carried out with the motion inputs obtained from the study of kinematics.

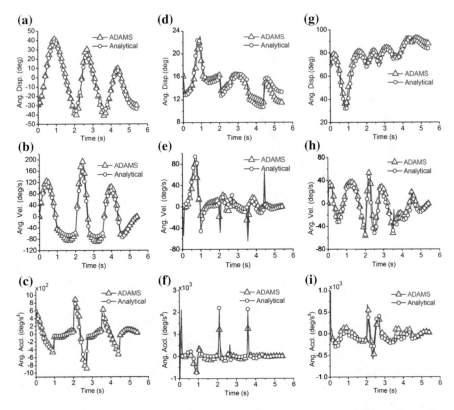

Fig. 5.12 Comparative graphs of the kinematic analysis of a realistic six-legged robot during turning motion on an uneven terrain using wave gait (DF = 1/2) for leg 6. Joint 11 **a** Angular displacement, **b** Angular velocity, **c** Angular acceleration. Joint 12 **d** Angular displacement, **e** Angular velocity, **f** Angular acceleration. Joint 13 **g** Angular displacement, **h** Angular velocity, **i** Angular acceleration

The frames G_0 and G are assumed to be parallel and coincident at origin $\mathbf{O}$, that is, $\boldsymbol{\eta}_G = (0, 0, 0)^T$. The robot's trunk body remains parallel to the slope of staircase at any instant of time. Therefore, initial position and orientation of $\mathbf{P}_0$ with respect to frame G are given by $\mathbf{p}_0^G = \{0, 0.412, 0.235, \tan^{-1}(h_s/b_s), 0, 0\}^T$. The corresponding initial joint angles are calculated from the CAD model and taken in the order of $[\theta_{i1}, \beta_{i2}, \beta_{i3}]$ (for $i = 1-6$), so that these angles become equal to $[5°, -16°, -69°]$ for leg 1, $[12°, 16°, 69°]$ for leg 2, $[12°, -16°, -69°]$ for leg 3, $[18.5°, 16°, 69°]$ for leg 4, $[22°, -16°, -69°]$ for leg 5, $[27.5°, 16°, 69°]$ for leg 6. The initial velocity of the trunk body moving parallel to the slope is assumed to be $v_{yz} = 0.1$ m/s. The maximum swing height of each leg along $\mathbf{Z}$ with respect to local frame of reference (L_{i3}) attached to leg tip at $\mathbf{P}_{i3}$ (start point of swing) became equal to $Hm_{in} + \Delta h = 0.072$ m, where Hm_{in} is the maximum height of the terrain on the path of swing leg i in the nth duty cycle, Δh is the additional clearance between terrain and swing height Hm_{in} (refer to [3]). The value of landing height with respect to frame L_{i3}, at the instant the leg comes to support is assumed to be

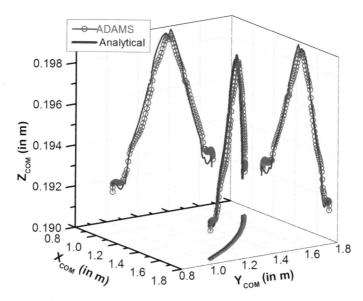

Fig. 5.13 Comparison of the analytical results of the aggregate COM of the system with respect to frame G_0 during turning motion on an uneven terrain with that of MSC.ADAMS®

equal to $h'_{in} = h_s = 0.05$ m. The maximum slip velocity (v_f) is assumed to be equal to 0.001 m/s with a slip angle of $\varepsilon_s = 45°$. The simulation operation is carried out in MATLAB for three duty cycles $(n = 3)$ with time step (h) of 0.05 s and trunk body's stroke length (s_0) of 0.125 m.

The elapsed time for three duty cycles is computed as 8.6 s with the time period of first, second, and third cycles as 3.0 s, 2.6 s, and 3.0 s, respectively. The computed kinematic motion parameters (displacement, velocity, acceleration, etc.), based on the motion and gait planning algorithms are provided as necessary inputs to the inverse dynamics model. Besides those parameters, the physical and contact parameters defined in Tables 3.1 and 4.2, respectively, are also given as inputs. Additionally, the actuator torque is limited to ±6 Nm for computation of optimal feet forces, joint torques, etc. The optimization algorithm chosen is *interior-point-convex quadprog* that satisfies the boundary conditions at each iteration corresponding to the objective function.

Subsequently, to validate the analytical model using VP tools, the 3D CAD model is preprocessed in CATIA SimDesigner and imported into MSC.ADAMS®, so that relevant inputs like joint angular velocities (results obtained analytically), contact parameters of the interacting surface, etc., could be defined for the CAD model before execution. In the present case, MSC.ADAMS® solver uses a default *gear stiff* (GSTIFF) *integrator* (backward differentiation formulation) algorithm to solve the DAEs of the dynamic problem [4]. Here, the time interval of 0.05 s has been kept the same as that obtained analytically for developing the simulation of the system. The computed results like the optimal force distribution in the legs, joint torques, power consumptions obtained by using (a) analytical method and (b) MSC.ADAMS® are analysed over three gait cycles keeping the total cycle time same. The results are in close agreement, which proves the efficacy of the developed algorithm.

Figure 5.14 shows the optimal distribution of feet forces $\left(F_{iz}^{G_0} \right)$ in all the legs of the six-legged robot plotted against time interval of $h = 0.05$ s. For the wave

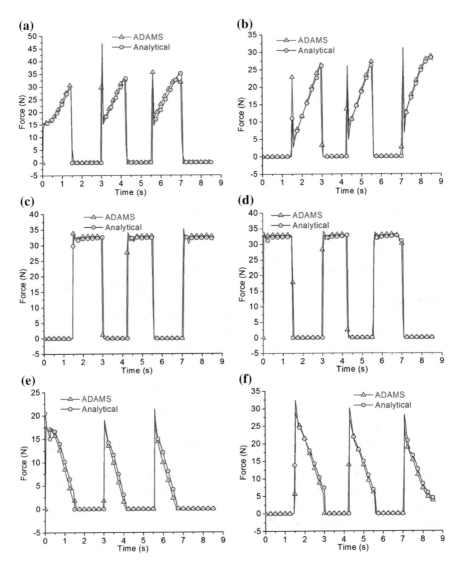

Fig. 5.14 Normal contact force distribution on the leg tips of the six-legged robot with respect to frame G_0 while ascending a staircase with tripod gait. **a** Leg 1, **b** leg 2, **c** leg 3, **d** leg 4, **e** leg 5, **f** leg 6

gait strategy with DF = 1/2, the support and swing phase times are kept equal for all the legs. It followed a wave sequence like (1) legs 1–4–5 in support phase and legs 2–3–6 in swing phase; (2) legs 2–3–6 in support phase and legs 1–4–5 in swing phase. The graphs reveal that the pattern of the normal reaction foot force $\left(F_{iz}^{G_0}\right)$ experienced by the legs 1–2, 3–4, and 5–6 are similar except that the pattern is out of phase by 180°. This can be attributed to the fact that wave gaits are regular and symmetric, with the right and left legs of each column having a phase difference of a half cycle. Also, it is seen that when a leg is in swing motion, the forces in that leg is zero and denoted by straight line. When the robot climbs upstairs, the entire COG of the system moves forward. This could be well explained by the graph, since the normal feet forces on the front legs in support phase increase, while that on the rear legs decrease with time till the start of next swing phase. Moreover, the variations of normal force distributions in the middle legs (refer to Fig. 5.14c, d) are found to be less and the values are approximately constant. This is attributed to the fact that the foothold positions of the legs during support phase are close to the COG of the system.

It is also interesting to note that summation of all the normal reaction feet forces at any instant of time with respect to G_0 balanced the weight of the six-legged robot with payload (i.e., 65.7 N) except during impact, which is assumed to last for the fraction of a second (refer to Fig. 5.15c). A momentary impulsive force is generated, when the leg tip collides with the terrain at the start of support phase. There is a sudden shoot in the normal reaction foot force $F_{iz}^{G_0}$, as can be well observed from Fig. 5.15c. It is also important to mention that the horizontal components of forces ($F_{ix}^{G_0}$ and $F_{iy}^{G_0}$) did exist, although their magnitude is not significant enough (Fig. 5.15a, b). Similarly, it is observed that the summation of the moment of forces acting at the

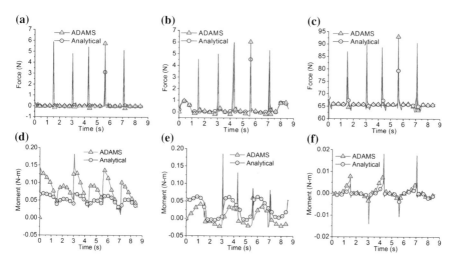

Fig. 5.15 Summation of forces and moments distribution with respect to frame G_0 **a** $\sum F_{ix}$, **b** $\sum F_{iy}$, **c** $\sum F_{iz}$, **d** $\sum T_{ix}$, **e** $\sum T_{iy}$, **f** $\sum T_{iz}$, where $i = 1$–6

foot-tips of support legs (refer to Fig. 5.15d, e) are insignificant and kept within the average range of +0.075 to −0.01 Nm (MATLAB) excluding the sudden shoots due to impact.

Figure 5.16 shows the torque distributions in all the joints of the legs. It is observed that the torque values during support phases are significantly higher compared to that

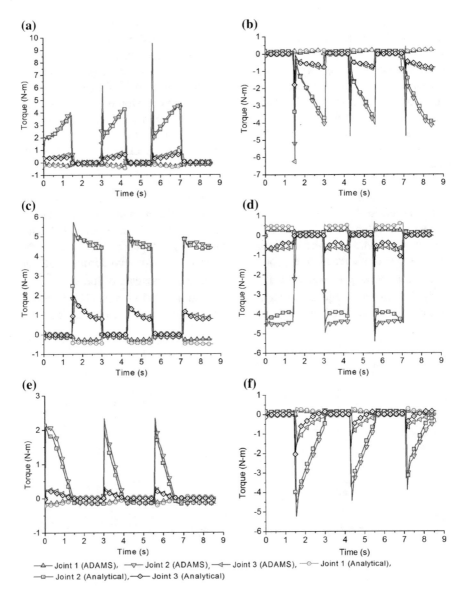

Fig. 5.16 Torque distribution in various joints of the six-legged robot while ascending a staircase with tripod gait. **a** Leg 1, **b** leg 2, **c** leg 3, **d** leg 4, **e** leg 5, **f** leg 6

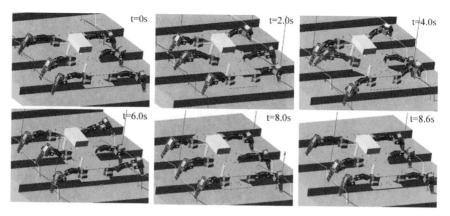

Fig. 5.17 Snapshots of a realistic six-legged robot simulated in MSC.ADAMS® while climbing a staircase with tripod gait

during swing phase for each leg. Also, the torque experienced by joint $i2$ of each leg is higher compared to that experienced by the other two joints at any instant of time. Moreover, torque variations in the middle legs are less compared to that of the front and rear legs (refer to Fig. 5.16c, d), since the location of the joints of these legs are close to the COG of the system. It is also interesting to note that the impact force (sudden high shoot out as shown in Fig. 5.14) experienced by any leg, when it touches the ground has significant effect on the joint torques (sudden high shoot out as shown in Fig. 5.16) of its own leg as well as considerable effect (small spikes as shown in Fig. 5.16) on the joints of other legs.

Figure 5.17 shows the snapshots of the simulations carried out in MSC.ADAMS® of a realistic six-legged robot ascending a staircase. From the motion of the robot, it is evident that the system is stable and followed a desired tripod gait sequence of 1–4–5 and 2–3–6.

5.2.2.2 Case Study 2: Experimentation with a Hex Crawler HDATS Robot Maneuvering on a Concrete Floor with Straight-Forward Motion

An attempt has been made to correlate the experimental results of a real six-legged robot with the simulated results of a realistic one without any payload for DF = 1/2. To stay as closely as possible to the reality, simulation environment has been kept the same, as it would be executed on real robot. The experimental results are compared with that of analytical results of the proposed model executed in MATLAB as well as that obtained by solving in MSC.ADAMS® subjected to the same initial inputs for all the three modes of measurement.

Experiment has been carried out with a commercial Hex Crawler HDATS Robot moving on a flat concrete floor with straight-forward motion to obtain feet forces'

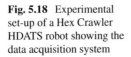

Fig. 5.18 Experimental
set-up of a Hex Crawler
HDATS robot showing the
data acquisition system

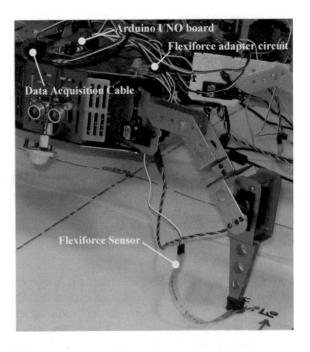

distributions data. It is conducted using a complete program installed on the robot
that could give command to the robot to perform straight-forward motion with DF =
1/2, trunk body stroke length (s_0) of 0.085 m, and trunk body velocity of 0.075 m/s.
The experiment is said to be completed after three duty cycles ($n = 3$). To measure
the feet forces, the head ends of the piezoresistive flexiforce sensors (six in numbers)
are attached to the footpad of each leg (refer to Fig. 5.18), while the conductive leads
at the tail end are attached to a flexiforce adapter circuit (filter cum amplifier circuit)
for obtaining steady noise-free force data. Each of the adapters is connected to an
Arduino UNO board (refer to Fig. 5.18), which is further connected to a remote
computer through com-port for real-time data acquisition. The data is filtered in
MATLAB for further processing. It is to be noted that the calibration of each of the
sensor is a must and is carried out using standard force measuring gauge. Further,
using three degree polynomial curve fit, a relationship is obtained between the sensor
reading and actual reading.

Simulation is carried out for the developed inverse dynamic model in MATLAB
for three duty cycles with the total time of analysis computed as 8.16 s (first cycle—
2.88 s, second cycle—2.4 s, third cycle—2.88 s) and time step of 0.08 s. Similarly,
computation is performed in MSC.ADAMS® solver using the CAD model keeping
all the conditions similar. The experimental data for the distribution of normal feet
forces are compared with the analytical and simulated data for over three duty cycles
with respect to G_0, as shown in Fig. 5.19. Looking at the plotted graphs of the normal
feet forces distribution in the different legs obtained by various methods (experiment,
analytical, and MSC.ADAMS®), it could be inferred that the distribution trends

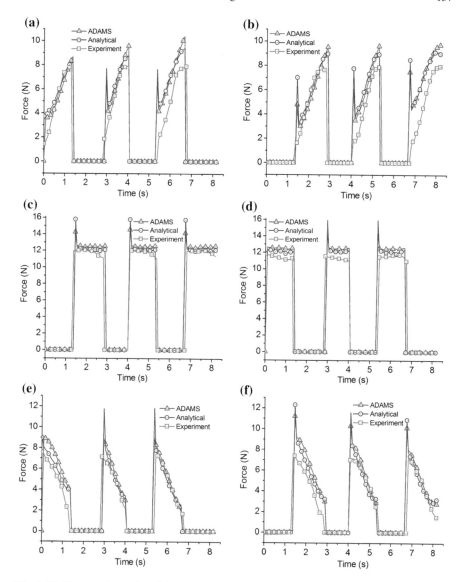

Fig. 5.19 Comparative study of the normal contact force distribution on the leg tips of the six-legged robot with respect to frame G_0 moving on a concrete floor with straight-forward motion and tripod gait. **a** Leg 1, **b** leg 2, **c** leg 3, **d** leg 4, **e** leg 5, **f** leg 6

are similar and in close agreement. It is also noticeable that though the developed mathematical model and the MSC.ADAMS® model of the realistic six-legged robot could generate an impact force instantaneously at the beginning of contact of feet tip with the ground during support phase, no such impact force has been observed during

Fig. 5.20 Comparative study of the summation of the normal contact force distribution, F_{iz} $(i = 1–6)$ with respect to frame G_0

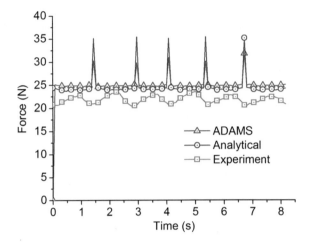

the experiment. The reason for such a behavior might be due to the fact that flexiforce sensors are not sensitive enough to sense the impact, which is instantaneous.

A comparative study of the summation of normal contact forces which should be equivalent to weight of the six-legged robot is illustrated in Fig. 5.20. At any instant, the experimental values are less compared to the computed ones. Moreover, it is also observed that for the three cycles, the sum of forces obtained experimentally varied in the range 20.0–23.5 N (approx.), while computationally the range is found to be from 24 to 25 N (approx.). Such variations in the experimental data are clearly due to the different portion of the area of leg tip of the Hex Crawler coming in contact with the ground during support phase, the most effective being at the time when the value of θ_{i1} is close to zero. This means that at every instant, the flexiforce sensor must be adjusted so that it comes into adequate ground contact, which is somewhat difficult with such type of sensor configuration. Such sensors should be customized to adapt the geometrical shape of the pad in contact with the ground, which is beyond the scope of the present book. But, overall the validation proves the efficacy of the proposed model with respect to trend followed by the normal force distribution.

Snapshots of the experimental (refer to Fig. 5.21) and simulated (refer to Fig. 5.22) data in MSC.ADAMS® show that the motion of the robot is stable and in accordance with the desired tripod gait sequence of 1–4–5 and 2–3–6.

5.2.2.3 Case Study 3: Experimentation with a Hex Crawler HDATS Robot Maneuvering on a Concrete Floor with Crab Motion Motion (DF = 1/2)

Like the previous case study (refer to Sect. 5.2.2.2), experimentation is also carried out with the available infrastructure on the Hex Crawler HDATS robot executing crab motion (DF = 1/2) on a flat concrete floor. Simulations have also been carried out in MSC.ADAMS® on a realistic six-legged robot that replicates the Hex Crawler,

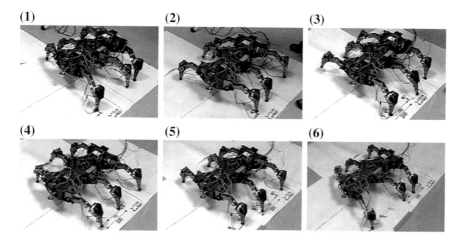

Fig. 5.21 Snapshots of a Hex Crawler HDATS robot moving on a concrete floor with tripod wave gait and straight-forward motion

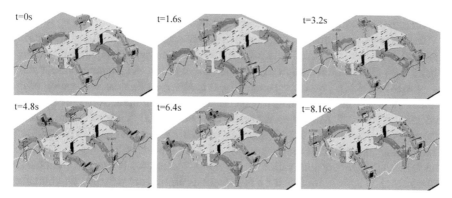

Fig. 5.22 Snapshots of a realistic six-legged robot simulated in MSC.ADAMS® moving on a concrete floor with tripod gait and straight-forward motion

keeping the environment same as that of the real experiment. Further, the developed analytical model is executed in MATLAB using the same boundary conditions as that of the realistic six-legged robot simulated in MSC.ADAMS®.

Experiment with the Hex Crawler has been conducted using reprogrammable code installed on the robot, such that it could perform crab motion with DF = 1/2, body stroke (s_0) of 0.025 m, and trunk body velocity of 0.010 m/s along X-direction maintaining a crab angle (θ_c) of 60° with respect to the longitudinal axis (Y-axis). The experiment is considered to be completed after three duty cycles ($n = 3$). Like the previous case study, data acquisition system arrangement is kept identical to measure the distributive feet forces (refer to Fig. 5.18). Calibration of the flexiforce sensors is done using standard gauge as discussed in the earlier section. The data are filtered in MATLAB for further processing. The same procedure has been followed as mentioned in Sect. 5.2.2.2.

Simulation is run in MATLAB using the developed inverse dynamic model for three duty cycles with the total time of executing the motion as 16.0 s (first cycle—5.5 s, second cycle—5.0 s, third cycle—5. 5 s) and time step of 0.05 s. Likewise, the CAD model with similar boundary conditions as that applied to the dynamic model is simulated using MSC.ADAMS$^{®}$ solver. The analytical data of the distribution of normal feet forces for three duty cycles are compared with the simulated data of MSC.ADAMS$^{®}$ and experimental data with respect to G_0 as shown in Fig. 5.23. A close look at the plots shows that the distribution of normal feet forces in the different legs obtained from experiments, simulation in MATLAB and simulation in MSC.ADAMS$^{®}$ follow a similar trend and are in close agreement. Further, the graphical plots in case of the analytical and MSC.ADAMS$^{®}$ models of the realistic six-legged robot shows sudden shoot outs (instantaneous force), which are due to collision of the leg tip with the ground. In case of experiment with Hex Crawler, no such shoot outs are observed in the plots though there is collision of the leg tip with the ground. This can be well explained with the fact that the flexiforce sensors are not sensitive enough to calculate the instantaneous impact.

The summation of all the normal reaction forces (excluding impact force during landing) equals to the total weight of the system at any instant during walking. Figure 5.24 gives an illustration of the summation of forces calculated analytically, using MSC.ADAMS$^{®}$ and experimentally. It is observed that the summation of the forces obtained from experiments is less compared to the computed ones in MATLAB and MSC.ADAMS$^{®}$. Moreover, it has also been observed that for the three cycles, the sum of forces obtained experimentally varied in the range 21.2–22.9 N (approx.), analytical results varied in the range 23.9–24.2 N (approx.) excluding impact force and MSC.ADAMS$^{®}$ results varied in the range 24.15–24.4 N (approx.) excluding impact force. Such variation of the experiment results in comparison with computed results might be well explained from the fact the force sensor attached to the leg pad is not having proper contact with the ground during the motion of the Hex Crawler. To minimize the error, such type of force sensors should be customized to adapt to the configuration of the portion of the pad in contact with the ground, but it is beyond the scope of the present book. Keeping aside, the validation proves the efficacy of the proposed model with respect to the distribution pattern of the normal force distribution.

Snapshots of the experimental (refer to Fig. 5.25) and simulated (refer to Fig. 5.26) data in MSC.ADAMS$^{®}$ display that the motion of the robot is stable and in accordance with the desired tripod gait sequence of 1–4–5 and 2–3–6.

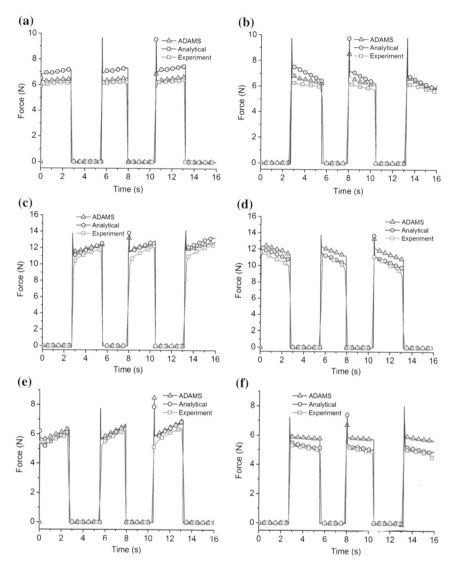

Fig. 5.23 Comparative study of the normal contact force distribution on the leg tips of the six-legged robot with respect to frame G_0 moving on a concrete floor with crab motion and tripod gait. **a** Leg 1, **b** leg 2, **c** leg 3, **d** leg 4, **e** leg 5, **f** leg 6

5.3 Summary

Both kinematic as well as dynamic analysis have been carried out analytically and their performances have been tested using available VP tools. A realistic CAD model has been developed, preprocessed with similar boundary constraints as the analytical

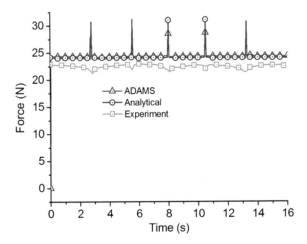

Fig. 5.24 Comparative study of the summation of the normal contact force distribution, F_{iz} ($i =$ 1–6) with respect to frame G_0 during crab motion

Fig. 5.25 Snapshots of a Hex Crawler HDATS robot moving on a concrete floor with tripod wave gait and crab motion

model, solved in a numerical solver (MSC.ADAMS) and further post-processed for visualization of the computed data. Computation of data of the kinematic motion parameters, distribution of the feet forces, joint torques in the legs under different boundary conditions for the analytical model are found to be in close agreement with the MSC.ADAMS data. The experiments carried out with the available infrastructure on a real robot of similar geometrical parameters and under suitable boundary conditions have also confirmed the effectiveness of the developed algorithm. Such validation paves the way for further design iterations with the developed algorithm,

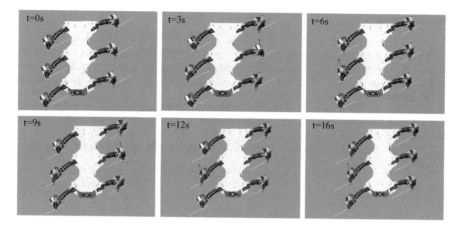

Fig. 5.26 Snapshots of a realistic six-legged robot simulated in MSC.ADAMS® moving on a concrete floor with tripod gait and crab motion

to improve the performance of the robot, thereby, aiming to find energy efficient model for six-legged robots under different surface conditions and gait strategies prior to the development of the first physical prototype.

Computer simulations results and subsequent validations using VP tools and experiments prove that the developed model is reliable and will be useful in the design and development of suitable control algorithms of six-legged robots. It is to be further noted that energy consumption and stability are two very important performance indices, which play key roles in the design and development of autonomous multi-legged robots. The parametric studies related to the effects of gait parameters on energy consumption and stability margins (both static and dynamic) will be useful in the selection of suitable gait parameters for a specific type of motion. The work carried out in the present work related to modeling, simulation, and subsequent validation of the complex kinematics and dynamics of a realistic six-legged robot maneuvering in various terrains, may be considered as one of the significant steps toward making such robots autonomous. However, for such autonomous robots, it is necessary to know input-output relationships of the governing parameters prior to execution of any specific task.

Further, for actual implementation of the developed algorithms in real robots, it is necessary to address the issues like online processing of the terrain information using sensors and/or cameras, design and development of suitable control scheme along with adaptive controller (must be able to provide controller gain values) and others. Hence, the developed model can serve as a tool for researchers, who would like to predict the performance of the legged robotic system during its locomotion on various terrains moving with straight-forward, crab and turning motion capabilities and adopting different gait strategies.

References

1. D.G. Ullman, *The Mechanical Design Process* (Mcgraw-Hill, New York, 1992)
2. M.J. Pratt, Virtual prototypes and product models in mechanical engineering, in *Virtual Prototyping: Virtual Environments and the Product Design Process*, ed. by J. Rix, S. Hass, J. Teixeira (Chapman and Hall, London, 1995)
3. A. Mahapatra, S.S. Roy, D.K. Pratihar, Computer Aided Modeling and Analysis of Turning Motion of Hexapod Robot on Varying Terrains. Int. J. Mechan. Mater. Des. **11**(3), 309–336 (2015)
4. C.W. Gear, Simultaneous numerical solution of differential-algebraic equations. IEEE Trans. Circuit Theory **18**(1), 89–95 (1971)

Appendix

Appendix A.1 Matrix Projectors

$$\mathbf{P}_r^T(x, y) = \begin{bmatrix} 1 & 0 & 0 \\ 0 & 1 & 0 \end{bmatrix} \tag{A.1}$$

$$\mathbf{P}_r^T(x, z) = \begin{bmatrix} 1 & 0 & 0 \\ 0 & 0 & 1 \end{bmatrix} \tag{A.2}$$

$$\mathbf{P}_r(y) = \begin{bmatrix} 0 & 1 & 0 \end{bmatrix}^T \tag{A.3}$$

$$\mathbf{P}_r(z) = \begin{bmatrix} 0 & 0 & 1 \end{bmatrix}^T \tag{A.4}$$

Appendix A.2 Loop Equations w.r.t Frame G

Each of the joints of the system has been described by geometric relations between frame G and/or relative (local) coordinates and body-fixed vectors and orientations. The geometric relations are included in the system model equations by means of suitably chosen projections and representations of *vector loop equations* and/or *orientation loop equations*. The appendix describes the derivation of the loop equations represented with respect to frame G located at origin $\mathbf{O}$ (refer to Fig. A.1a). L_0 is the body-fixed reference frame located at point $\mathbf{P}_0$ on the trunk body. L_i' and L_i'' are the local frames of the successive joint states respectively located at $\mathbf{S}_i$ as shown in Fig. A.1b. Similarly, the local frames L_{ij}' and L_{ij}'' are located at P_{ij} where $j = 1$ to 2 (refer Fig. A.1b). L_{i3} and L'_{i3} are the tip-point reference frames located at $\mathbf{P}_{i3}$ (refer to Fig. 3.3b). Also, $\mathbf{r}_{P_0 O}^G$, $\mathbf{r}_{S_i O}^G$, and $\mathbf{r}_{P_{ij} O}^G$ ($j = 1$ to 3) are the displacement vectors from point $\mathbf{O}$ to $\mathbf{P}_0$, $\mathbf{S}_i$, and $\mathbf{P}_{ij}$, respectively, represented in global reference frame

© Springer Nature Singapore Pte Ltd. 2020
A. Mahapatra et al., *Multi-body Dynamic Modeling of Multi-legged Robots*,
Cognitive Intelligence and Robotics, https://doi.org/10.1007/978-981-15-2953-5

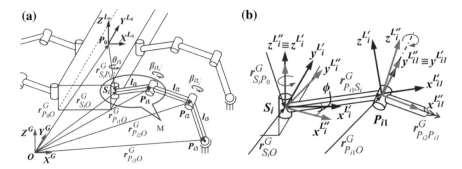

Fig. A.1 Reference frames and vector loops of the robotic system: **a** isometric view, **b** view 'M'

G. Here, ϕ is twisted angle of the coxa (in the present study $\phi = 0$). Loop equations (both vector loop and orientation loop) of the system are derived as given below:

1. **Loop OP_0S_iO**

Joint 'J_{i1}' (refer to Fig. 3.1a) is considered to be a revolute joint, with rotation about z-axis by 'θ_{i1}'. The constraint equations are,

(a) *Vector Loop Equation w.r.t.* G

With respect to frame (G), the vector loop equation can be written as,

$$\mathbf{r}_{P_0 O}^{G} + \mathbf{r}_{S_i P_0}^{G} - \mathbf{r}_{S_i O}^{G} = \mathbf{0}_3 \tag{A.5}$$

or

$$\mathbf{r}_{P_0 O}^{G} + \mathbf{A}^{GL_0}\mathbf{r}_{S_i P_0}^{G} - \mathbf{r}_{S_i O}^{G} = \mathbf{0}_3 \tag{A.6}$$

(b) *Orientation Loop Equation w.r.t.* G

$$\mathbf{A}^{GL_0}.\mathbf{A}^{L_0 G} = \mathrm{I}_3 \tag{A.7}$$

or

$$\underbrace{\mathbf{A}^{GL_i''}}_{\substack{free \\ rotation}} . \underbrace{\mathbf{A}^{L_i'' L_i'}}_{\substack{free\ relative \\ rotation}} . \underbrace{\mathbf{A}^{L_i' L_0}}_{\substack{fixed \\ rotation}} . \underbrace{\mathbf{A}^{L_0 G}}_{\substack{free \\ rotation}} = \mathrm{I}_3 \tag{A.8}$$

or

$$\mathbf{A}^{GL_i''} = \mathbf{A}^{GL_0}.\mathbf{A}^{L_0 L_i'}.\mathbf{A}^{L_i' L_i''} \tag{A.9}$$

NB:

Transformation of frame from G to L_0,

$$\mathbf{A}^{L_0 G} = \begin{bmatrix} c_2 c_3 & c_1 s_3 + s_1 s_2 c_3 & s_1 s_3 - c_1 s_2 c_3 \\ -c_2 s_3 & c_1 c_3 - s_1 s_2 s_3 & s_1 c_3 + c_1 s_2 s_3 \\ s_2 & -s_1 c_2 & c_1 c_2 \end{bmatrix}$$

where $c_1 = \cos \alpha_0$, $s_1 = \sin \alpha_0$, $c_2 = \cos \beta_0$, $s_2 = \sin \beta_0$, $c_3 = \cos \theta_0$, $s_3 = \sin \theta_0$.

Using *orthogonal property* of the matrix we have,

$$\mathbf{A}^{GL_0} = (\mathbf{A}^{L_0 G})^{-1} = (\mathbf{A}^{L_0 G})^T$$

Transformation of frame from L'_i to L''_i,

$$\underbrace{\mathbf{A}^{L''_i L'_i}}_{L'_i \to L''_i} = \begin{bmatrix} c_{i1} & s_{i1} & 0 \\ -s_{i1} & c_{i1} & 0 \\ 0 & 0 & 1 \end{bmatrix} \text{(CCW about z-axis)}$$

where $c_{i1} = \cos \theta_{i1}$, $s_{i1} = \sin \theta_{i1}$

Using *orthogonal property* of the matrix we have,

$$\mathbf{A}^{L'_i L''_i} = (\mathbf{A}^{L''_i L'_i})^{-1} = (\mathbf{A}^{L''_i L'_i})^T$$

2. **Loop $OS_i P_{i1} O$**

Joint 'J_{i2}' (refer to Fig. 3.1a) is considered to be a revolute joint with rotation about y-axis by 'β'_{i2}. Therefore,

(a) *Vector Loop Equation w.r.t. G*

With respect to frame (G), the vector loop equation can be written as,

$$\mathbf{r}^G_{S_i O} + \mathbf{r}^G_{P_{i1} S_i} - \mathbf{r}^G_{P_{i1} O} = \mathbf{0}_3 \qquad (A.10)$$

or

$$\mathbf{r}^G_{S_i O} + \mathbf{A}^{GL''_i} \mathbf{r}^{L''_i}_{P_{i1} S_i} - \mathbf{r}^G_{P_{i1} O} = \mathbf{0}_3 \qquad (A.11)$$

(b) *Orientation Loop Equation w.r.t. G*

$$\mathbf{A}^{GL''_i} . \mathbf{A}^{L''_i G} = I_3 \qquad (A.12)$$

or

$$\underbrace{\mathbf{A}^{GL_{i1}''}}_{\substack{free \\ rotation}} \cdot \underbrace{\mathbf{A}^{L_{i1}''L_{i1}'}}_{\substack{free \\ rel.vel.}} \cdot \underbrace{\mathbf{A}^{L_{i1}'L_i''}}_{\substack{fixed \\ rotation}} \cdot \underbrace{\mathbf{A}^{L_i''G}}_{\substack{free \\ rotation}} = \mathbf{I}_3 \tag{A.13}$$

or

$$\mathbf{A}^{GL_{i1}''} = \mathbf{A}^{GL_i''}.\mathbf{A}^{L_i''L_{i1}'}.\mathbf{A}^{L_{i1}'L_{i1}''} \tag{A.14}$$

NB:

Transformation of frame from $\mathbf{L}_0$ to L_i',

$$\underbrace{\mathbf{A}^{L_i'L_0}}_{L_0 \to L_i'} = \begin{bmatrix} c_\gamma & -s_\gamma & 0 \\ s_\gamma & c_\gamma & 0 \\ 0 & 0 & 1 \end{bmatrix} \text{(CW about z-axis)}$$

where $c_\gamma = \cos\gamma$, $s_\gamma = \sin\gamma$, γ being the angle between frames L_i' and $\mathbf{L}_0$. Using *orthogonal property* of the matrix we have,

$$\mathbf{A}^{L_0 L_i'} = (\mathbf{A}^{L_i'L_0})^{-1} = (\mathbf{A}^{L_i'L_0})^T$$

Transformation of frame from Local ($\mathbf{L'}_{i1}$) to Local (L_{i1}''),

$$\underbrace{\mathbf{A}^{L_{i1}''L_{i1}'}}_{L_{i1}' \to L_{i1}''} = \begin{bmatrix} c_{i2} & 0 & -s_{i2} \\ 0 & 1 & 0 \\ s_{i2} & 0 & c_{i2} \end{bmatrix} \text{(CCW about y-axis)}$$

where $c_{i2} = \cos\beta_{i2}$, $s_{i2} = \sin\beta_{i2}$.
Using *orthogonal property* of the matrix we have,

$$\mathbf{A}^{L_{i1}'L_{i1}''} = (\mathbf{A}^{L_{i1}''L_{i1}'})^{-1} = (\mathbf{A}^{L_{i1}''L_{i1}'})^T$$

---

3. Loop $\mathbf{OP}_{i1}\mathbf{P}_{i2}\mathbf{O}$

Joint 'J_{i3}' (refer to Fig. 3.1a) is considered to be a revolute joint, with rotation about y-axis as $'\beta_{i3}'$

(a) *Vector Loop Equation w.r.t.* $\mathbf{G}$

With respect to frame ($\mathbf{G}$), the vector loop equation can be written as,

$$\mathbf{r}_{P_{i1}O}^G + \mathbf{r}_{P_{i2}P_{i1}}^G - \mathbf{r}_{P_{i2}O}^G = \mathbf{0}_3 \tag{A.15}$$

or

$$\mathbf{r}^{G}_{P_{i1}O} + \mathbf{A}^{GL''_{i1}}\mathbf{r}^{L''_{i1}}_{P_{i2}P_{i1}} - \mathbf{r}^{G}_{P_{i2}O} = \mathbf{0}_3 \tag{A.16}$$

(b) *Orientation Loop Equation w.r.t.* **G**

$$\mathbf{A}^{GL''_{i1}}.\mathbf{A}^{L''_{i1}G} = \mathbf{I}_3 \tag{A.17}$$

or

$$\underbrace{\mathbf{A}^{GL''_{i2}}}_{\substack{free\\rotation}} \cdot \underbrace{\mathbf{A}^{L''_{i2}L'_{i2}}}_{\substack{free\;relative\\rotation}} \cdot \underbrace{\mathbf{A}^{L'_{i2}L''_{i1}}}_{\substack{fixed\\rotation}} \cdot \underbrace{\mathbf{A}^{L''_{i1}G}}_{\substack{free\\rotation}} = \mathbf{I}_3 \tag{A.18}$$

or

$$\mathbf{A}^{GL''_{i2}} = \mathbf{A}^{GL''_{i1}}.\mathbf{A}^{L''_{i1}L'_{i2}}.\mathbf{A}^{L'_{i2}L''_{i2}} \tag{A.19}$$

NB:
   Transformation of frame from L''_i to L'_{i1},

$$\underbrace{\mathbf{A}^{L'_{i1}L''_i}}_{L''_i \rightarrow L'_{i1}} = \begin{bmatrix} c_\phi & 0 & s_\phi \\ 0 & 1 & 0 \\ -s_\phi & 0 & c_\phi \end{bmatrix} \text{(CW about y-axis)}$$

where $c_\phi = \cos\phi$, $s_\phi = \sin\phi$, ϕ being the angle between frames L''_i and L'_{i1}.
Using *orthogonal property* of the matrix we have,

$$\mathbf{A}^{L''_i L'_{i1}} = (\mathbf{A}^{L'_{i1}L''_i})^{-1} = (\mathbf{A}^{L'_{i1}L''_i})^T$$

Transformation of frame from local (L'_{i2}) to local (L''_{i2}),

$$\underbrace{\mathbf{A}^{L''_{i2}L'_{i2}}}_{L'_{i2} \rightarrow L''_{i2}} = \begin{bmatrix} c_{i3} & 0 & -s_{i3} \\ 0 & 1 & 0 \\ s_{i3} & 0 & c_{i3} \end{bmatrix} \text{(CCW about y-axis) where } c_{i3} = \cos\beta_{i3}, \; s_{i3} = \sin\beta_{i3}.$$

Using *orthogonal property* of the matrix we have,

$$\mathbf{A}^{L'_{i2}L''_{i2}} = (\mathbf{A}^{L''_{i2}L'_{i2}})^{-1} = (\mathbf{A}^{L''_{i2}L'_{i2}})^T$$

4. Loop $OP_{i2}P_{i3}O$

Joint 'J_{i4}' is considered to be a *spherical joint* between link '$i3$' and ground.

(a) *Vector Loop Equation w.r.t.* **G**

$$\mathbf{r}^G_{P_{i2}O} + \mathbf{r}^G_{P_{i3}P_{i2}} - \mathbf{r}^G_{P_{i3}O} = \mathbf{0}_3 \tag{A.20}$$

or

$$\mathbf{r}^G_{P_{i2}O} + \mathbf{A}^{GL''_{i2}}\mathbf{r}^{L''_{i2}}_{P_{i3}P_{i2}} - \mathbf{r}^G_{P_{i3}O} = \mathbf{0}_3 \tag{A.21}$$

(b) Constraint position equation if '$\mathbf{r}^G_i$' is a point on the planned trajectory,

$$\mathbf{r}^G_{P_{i3}O} - \mathbf{r}^G_i = \mathbf{0}_3 \tag{A.22}$$

Appendix A.3 Important Transformation Matrices

$$\mathbf{A}^{GG_0} = \begin{bmatrix} \cos\beta_G\cos\theta_G & \cos\alpha_G\sin\theta_G + \sin\alpha_G\sin\beta_G\cos\theta_G & \sin\alpha_G\sin\theta_G - \cos\alpha_G\sin\beta_G\cos\theta_G \\ -\cos\beta_G\sin\theta_G & \cos\alpha_G\cos\theta_G - \sin\alpha_G\sin\beta_G\sin\theta_G & \sin\alpha_G\cos\theta_G + \cos\alpha_G\sin\beta_G\sin\theta_G \\ \sin\beta_G & -\sin\alpha_G\cos\beta_G & \cos\alpha_G\cos\beta_G \end{bmatrix} \tag{A.23}$$

$$\mathbf{A}^{L_0G} = \begin{bmatrix} \cos\beta_0\cos\theta_0 & \cos\alpha_0\sin\theta_0 + \sin\alpha_0\sin\beta_0\cos\theta_0 & \sin\alpha_0\sin\theta_0 - \cos\alpha_0\sin\beta_0\cos\theta_0 \\ -\cos\beta_0\sin\theta_0 & \cos\alpha_0\cos\theta_0 - \sin\alpha_0\sin\beta_0\sin\theta_0 & \sin\alpha_0\cos\theta_0 + \cos\alpha_0\sin\beta_0\sin\theta_0 \\ \sin\beta_0 & -\sin\alpha_0\cos\beta_0 & \cos\alpha_0\cos\beta_0 \end{bmatrix} \tag{A.24}$$

$$\mathbf{A}^{G_0G} = (\mathbf{A}^{GG_0})^{-1} = (\mathbf{A}^{GG_0})^T \tag{A.25}$$

$$\mathbf{A}^{GL_0} = (\mathbf{A}^{L_0G})^{-1} = (\mathbf{A}^{L_0G})^T \tag{A.26}$$

$$\mathbf{A}^{G_0L_0} = \mathbf{A}^{G_0G}.\mathbf{A}^{GL_0} \tag{A.27}$$

$$\mathbf{A}^{GL_k} = \mathbf{A}^{GL_0}.\mathbf{A}^{L_0L_k} \tag{A.28}$$

$$\mathbf{A}^{L_0L_k} = \begin{bmatrix} \cos(\gamma-\theta_{i1})\cos\lambda_1 & \sin(\gamma-\theta_{i1}) & -\cos(\gamma-\theta_{i1})\sin\lambda_1 \\ -\sin(\gamma-\theta_{i1})\cos\lambda_1 & \cos(\gamma-\theta_{i1}) & \sin(\gamma-\theta_{i1})\sin\lambda_1 \\ \sin\lambda_1 & 0 & \cos\lambda_1 \end{bmatrix} \tag{A.29}$$

where

(i) If $L_k = L''_i = L'_{i1}$, then $\lambda_1 = 0$, (L''_i and L'_{i1} are parallel frames)

(ii)   If $L_k = L''_{i1} = L'_{i2}$, then $\lambda_1 = \phi - \beta_{i2}$ (L''_{i1} and L'_{i2} are parallel frames)

(iii) If $L_k = L''_{i2} = L'_{i3}$, then $\lambda_1 = \phi - \beta_{i2} - \beta_{i3}$ (L''_{i2} and L'_{i3} are parallel frames).

$$\mathbf{A}^{L_{i3}L_k} = \begin{bmatrix} \sin\lambda_2 & \cos\lambda_2 & 0 \\ -\cos\lambda_2 & \sin\lambda_2 & 0 \\ 0 & 0 & 1 \end{bmatrix} \tag{A.30}$$

where

(i)   If $L_k = L''_{i3}$, then $\lambda_2 = \theta_c$,

(ii)   If $L_k = L'''_{i3}$, then $\lambda_2 = \theta_{i3}^{L_{i3}}$.

Here, θ_c and $\theta_{i3}^{L_{i3}}$ are the crab and turning angles executed during crab and turning motion of the robot, respectively.

$$\mathbf{A}^{G_0 N_{i3}} = \mathbf{A}^{G_0 G} . \mathbf{A}^{G N_{i3}} \tag{A.31}$$

$$\mathbf{A}^{G N_{i3}} = \mathbf{I}_3 (\text{since, } G \text{ and } N_{i3} \text{ are assumed parallel}) \tag{A.32}$$

where $\mathbf{I}_3$ is an identity matrix.

Appendix A.4 Trajectory Planning of Swing Leg

Trajectory planning of tip point $\mathbf{P}^s_{i3}$ of the swing is carried in 3D Cartesian space (refer to Fig. 3.4).

I. *Straight-Forward and Turning Motion*

During straight-forward and turning motion, the swing leg follows a 3D trajectory in the Cartesian space with respect to local frames $\mathbf{L}_{i3}$ and $\mathbf{L}'''_{i3}$, respectively. The mathematical calculations in sections (a) and (b) are for the points $\mathbf{P}_{i3}$, $\mathbf{Q}_{i3}$, $\mathbf{T}_{i3}$, and $\mathbf{P}'_{i3}$ located in the trajectory of swing leg i and measured with respect to local frame $\mathbf{L}_{i3}$ during *straight-forward motion*. Refer to Fig. 3.4a, c.

(a)   **Calculation of coordinates of points $\mathbf{P}_{i3}$, $\mathbf{Q}_{i3}$, $\mathbf{T}_{i3}$, and $\mathbf{P}'_{i3}$ with respect to frame L_{i3}**

   Coordinates of $\mathbf{P}_{i3}$

$$x_{P_{i3}}^{L_{i3}} = 0; \tag{A.33}$$

$$y_{P_{i3}}^{L_{i3}} = 0; \tag{A.34}$$

$$z^{L_{i3}}_{P_{i3}} = 0 \tag{A.35}$$

Coordinates of Q_{i3}

$$x^{L_{i3}}_{Q_{i3}} = x^{L_{i3}}_{P_{i3}} + \Delta z_{P_{i3}Q_{i3}}. \tan \gamma_{xz} \tag{A.36}$$

$$y^{L_{i3}}_{Q_{i3}} = y^{L_{i3}}_{P_{i3}} + \Delta z_{P_{i3}Q_{i3}}. \tan \gamma_{yz} \tag{A.37}$$

$$z^{L_{i3}}_{Q_{i3}} = H_{m_{i1}} + \Delta h \tag{A.38}$$

where

$$\Delta z_{P_{i3}Q_{i3}} = z^{L_{i3}}_{Q_{i3}} - z^{L_{i3}}_{P_{i3}} = H_{m_{i1}} + \Delta h \tag{A.39}$$

Coordinates of T_{i3}

$$x^{L_{i3}}_{T_{i3}} = x^{L_{i3}}_{P'_{i3}} + \Delta z_{P'_{i3}T_{i3}}. \tan \gamma'_{xz} \tag{A.40}$$

$$y^{L_{i3}}_{T_{i3}} = y^{L_{i3}}_{P'_{i3}} - \Delta z_{P'_{i3}T_{i3}}. \tan \gamma'_{xz} \tag{A.41}$$

$$z^{L_{i3}}_{T_{i3}} = z^{L_{i3}}_{Q_{i3}} = z^{L_{i3}}_{R_{i3}} = H_{m_{i1}} + \Delta h \tag{A.42}$$

where

$$\Delta z_{P'_{i3}T_{i3}} = z^{L_{i3}}_{T_{i3}} - z^{L_{i3}}_{P'_{i3}} = H_{m_{i1}} + \Delta h - h'_{i3} \tag{A.43}$$

$(\gamma_{xz}, \gamma_{yz})$ are the angle of ascend and $(\gamma'_{xz}, \gamma'_{yz})$ are the angle of descend of the trajectory in XZ and YZ plane, respectively.
Coordinates of P'_{i3}

$$x^{L_{i3}}_{P'_{i3}} = x^{L_{i3}}_{P_{i3}} = 0; \tag{A.44}$$

$$y^{L_{i3}}_{P'_{i3}} = y^{L_{i3}}_{P_{i3}} + s^f_w \tag{A.45}$$

$$z^{L_{i3}}_{P'_{i3}} = h'_{i3} \tag{A.46}$$

Here, s^f_w designates the full stroke of swing leg during the turning motion. It is to be noted that for turning motion, the calculations will be with respect to local frame L'''_{i3}.

(b) Swing leg trajectory planning with respect to frame L_{i3}

Along X-axis:

$$
\begin{aligned}
x_{ei} &= a_{i0} + a_{i1}.t + a_{i2}.t^2/2 + a_{i3}.t^3/3 && (\mathbf{P}_{i3} \text{ to } \mathbf{Q}_{i3}) \\
&= b_{i0} + b_{i1}t + b_{i2}t^2 + b_{i3}t^3 + b_{i4}t^4 + b_{i5}t^5 && (\mathbf{Q}_{i3} \text{ to } \mathbf{T}_{i3}) \\
&= c_{i0} + c_{i1}.t + c_{i2}.t^2/2 + c_{i3}.t^3/3 && \left(\mathbf{T}_{i3} \text{ to } \mathbf{P}'_{i3}\right)
\end{aligned} \tag{A.47}
$$

The fourteen coefficients a_{i0}, a_{i1}, a_{i2} … c_{i3} and so on are computed using suitable boundary conditions and are given in Table A.1.

Along Y-axis ($\mathbf{P}_{i3}$ to $\mathbf{P}'_{i3}$):

$$
y_{ei} = y^{L_{i3}}_{P_{i3}} + a_{y_i^{PP'}}.\Delta^2_{PP'}.(3 - 2\Delta_{PP'}) \tag{A.48}
$$

where

$$
a_{y_i^{PP'}} = y^{L_{i3}}_{P'_{i3}} - y^{L_{i3}}_{P_{i3}} \tag{A.49}
$$

$$
\Delta_{PP'} = \left(t - t^s_{start}\right)/\left(t^s_{end} - t^s_{start}\right) \tag{A.50}
$$

$$
d\Delta_{PP'}/dt = 1/\left(t^s_{end} - t^s_{start}\right) \tag{A.51}
$$

Table A.1 Boundary conditions

	Time	Equation no.	Condition
BCs (t_0 to t_1)	At $t = t^s_{start}$,	(1)	$x_{ei} = x_{ei}\|_{t=t^s_{start}} = x_{P_{i3}}$
		(2)	$\dot{x}_{ei} = \dot{x}_{ei}\|_{t=t^s_{start}} = 0$
		(3)	$\ddot{x}_{ei} = \ddot{x}_{ei}\|_{t=t^s_{start}} = 0$
BCs (t_1 to t_2)	At $t = t^s_1$,	(4)	$x_{ei} = x_{ei}\|_{t=t^s_1} = x_{Q_{i3}}$
		(5)	$\dot{x}_{ei} = \dot{x}_{ei}\|_{t=t^s_1} = \dot{x}_{Q_{i3}}$
		(6)	$\ddot{x}_{ei} = \ddot{x}_{ei}\|_{t=t^s_1} = \ddot{x}_{Q_{i3}}$
	At $t = t^s_m$,	(7)	$\dot{x}_{ei} = \dot{x}_{ei}\|_{t=t^s_m} = \dot{x}_{R_{i3}} = 0$
	At $t = t^s_2$,	(8)	$x_{ei} = x_{ei}\|_{t=t^s_2} = x_{T_{i3}}$
		(9)	$\dot{x}_{ei} = \dot{x}_{ei}\|_{t=t^s_2} = \dot{x}_{T_{i3}}$
		(10)	$\ddot{x}_{ei} = \ddot{x}_{ei}\|_{t=t^s_2} = \ddot{x}_{T_{i3}}$
BCs (t_2 to t_3)	At $t = t^s_{end}$,	(11)	$x_{ei} = x_{ei}\|_{t=t^s_{end}} = x_{P'_{i3}}$
		(12)	$\dot{x}_{ei} = \dot{x}_{ei}\|_{t=t^s_{end}} = 0$
		(13)	$\ddot{x}_{ei} = \ddot{x}_{ei}\|_{t=t^s_{end}} = 0$
Addl BC		(14)	$x_{ei}\|_{t=t^s_{start}} = x_{ei}\|_{t=t^s_{end}}$ i.e. $x_{P_{i3}} = x_{P'_{i3}}$

Along Z-axis:

$$z_{ei} = z_{P_{i3}}^{L_{i3}} + a_{z_i PQ}.\Delta_{PQ}^2.(3 - 2\Delta_{PQ}) \quad (\mathbf{P}_{i3} \text{ to } \mathbf{Q}_{i3})$$

$$= z_{Q_{i3}}^{L_{i3}} \quad\quad\quad\quad\quad\quad\quad\quad\quad (\mathbf{Q}_{i3} \text{ to } \mathbf{T}_{i3})$$

$$= z_{T_{i3}}^{L_{i3}} - a_{z_i TP'}.\Delta_{TP'}^2.(3 - 2\Delta_{TP'}) \quad (\mathbf{T}_{i3} \text{ to } \mathbf{P}'_{i3}) \quad\quad\text{(A.52)}$$

where

$$a_{z_i PQ} = z_{Q_{i3}}^{L_{i3}} - z_{P_{i3}}^{L_{i3}} \quad\quad\quad\quad\quad\quad\quad\quad\quad\text{(A.53)}$$

$$a_{z_i TP'} = z_{T_{i3}}^{L_{i3}} - z_{P'_{i3}}^{L_{i3}} \quad\quad\quad\quad\quad\quad\quad\quad\text{(A.54)}$$

$$\Delta_{PQ} = \left(t - t_{start}^s\right)/\left(t_1^s - t_{start}^s\right) \quad\quad\quad\quad\quad\text{(A.55)}$$

$$\Delta_{TP'} = \left(t - t_2^s\right)/\left(t_{end}^s - t_2^s\right) \quad\quad\quad\quad\quad\quad\text{(A.56)}$$

Superscript 's' indicates swing. For every swing phase of leg i, t_{start}^s, t_1^s, t_2^s, and t_{end}^s indicate the start (point $\mathbf{P}_{i3}$), end of acceleration (point $\mathbf{Q}_{i3}$), start of deceleration (point $\mathbf{T}_{i3}$), and end (point $\mathbf{P}'_{i3}$) time, respectively.

Therefore, the coordinates of vector $\mathbf{r}_{P_{i3}^s P_{i3}}^{L_{i3}}$ (refer to Eq. (3.81)) are given by,

$$\mathbf{r}_{P_{i3}^s P_{i3}}^{L_{i3}} = (x_{ei}, y_{ei}, z_{ei})^T \quad\quad\quad\quad\quad\quad\text{(A.57)}$$

It is to be noted that for *turning motion*, the calculation procedure will be the same for measurement of the coordinates with respect to local frame $\mathbf{L}_{i3}'''$ such that the coordinates of vector $\mathbf{r}_{P_{i3}^s P_{i3}}^{L_{i3}'''}$ (refer to Eq. (3.83)) are given by,

$$\mathbf{r}_{P_{i3}^s P_{i3}}^{L_{i3}'''} = (x_{ei}, y_{ei}, z_{ei})^T \quad\quad\quad\quad\quad\quad\text{(A.58)}$$

(a) Calculation of turning angle $\theta_{i3}^{L_{i3}}$ ($i = 1$ to 6) during turning motion

In the present study, the turning angle $\theta_{i3}^{L_{i3}}$ has been introduced to execute the turning motion of the robot. Figure A.2 gives the kinematic scheme (top view) of the robot during turning motion with tripod wave gait. Some trigonometrical relations are established with respect to frame G, and the turning angle is calculated and generalized for other duty factors and gait strategy.

(i) Angular displacement per stroke of the trunk body,

$$\Delta\theta_0^c = s_0^c/\rho_0 \qu\quad\quad\quad\quad\quad\quad\quad\quad\text{(A.59)}$$

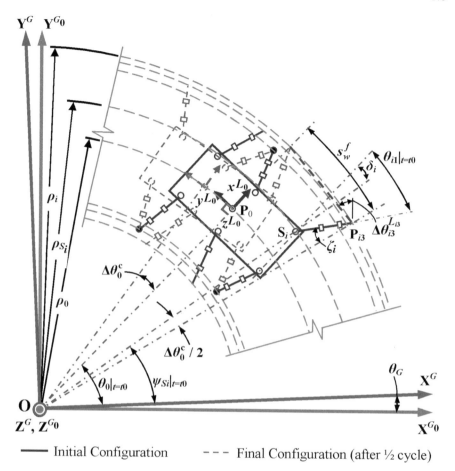

Fig. A.2 A kinematic scheme of the robot (top view) during turning motion with tripod wave gait (only one-half cycle)

$$s_0^c = m s_0''$$
(A.60)

where m is the number of divisions of a gait cycle

(ii) Angular stroke of the swing leg (full swing),

$$s_w^f = 2.\rho_i \cdot \sin(\Delta\theta_0^c/2)$$
(A.61)

(iii) $\angle S_i P_{i3} O = \sin^{-1}\left(\left(\rho_{S_i} \sin\left(K_1.\Delta\theta_0^c/m\right)\right)/l_i\right)$
(A.62)

(iv) $\zeta_i = \pi - \angle S_i P_{i3} O - K_1 . \Delta\theta_0^c / m$ (for interior legs $(i = 1, 3, 5)$ (A.63)

$= \angle S_i P_{i3} O + K_1 . \Delta\theta_0^c / m$ (for exterior legs $(i = 2, 4, 6)$)   (A.64)

(v) $\delta_i = \tan^{-1}\left(\dfrac{m_{S_2 S_1} - m_{S_i O}}{1 + m_{S_2 S_1} . m_{S_i O}}\right)$ for $i = 1$ to 2

$= \tan^{-1}\left(\dfrac{m_{S_4 S_3} - m_{S_i O}}{1 + m_{S_4 S_3} . m_{S_i O}}\right)$ for $i = 3$ to 4

$= \tan^{-1}\left(\dfrac{m_{S_6 S_5} - m_{S_i O}}{1 + m_{S_6 S_5} . m_{S_i O}}\right)$ for $i = 5$ to 6 (A.65)

(vi) Slopes:

$$m_{S_i O} = y_{S_i O}^G / x_{S_i O}^G \quad \text{for } i = 1 \text{ to } 6 \tag{A.66}$$

$$m_{S_2 S_1} = (y_{S_2 O}^G - y_{S_1 O}^G) / (x_{S_2 O}^G - x_{S_1 O}^G) \tag{A.67}$$

$$m_{S_4 S_3} = (y_{S_4 O}^G - y_{S_3 O}^G) / (x_{S_4 O}^G - x_{S_3 O}^G) \tag{A.68}$$

$$m_{S_6 S_5} = (y_{S_6 O}^G - y_{S_5 O}^G) / (x_{S_6 O}^G - x_{S_5 O}^G) \tag{A.69}$$

(vii) $\theta_{i1}|_{t=t_0} = \zeta_i - \delta_i$ (A.70)

The turning foot-tip radius (ρ_i) and the turning radius of joint $i1$ of leg 'i' (ρ_{S_i}) are calculated from the forward kinematics of the system as discussed in Sect. 3.1, such that,

$$\rho_i = \sqrt{\left(x_{P_{i3} O}^G\right)^2 + \left(y_{P_{i3} O}^G\right)^2} \tag{A.71}$$

$$\rho_{S_i} = \sqrt{\left(x_{S_i O}^G\right)^2 + \left(y_{S_i O}^G\right)^2} \tag{A.72}$$

Therefore, the turning angle $\theta_{i3}^{L_{i3}}$ $(i = 1$ to $6)$ for any gait configuration with respect to frame G is given by the relation:

$$\left.\theta_{i3}^{L_{i3}}\right|_n = \left.\psi_{S_i}\right|_{t=t_0|_1} + \left[6(n-1) + m_2 + m_1/2\right]\Delta\theta_0^c / m \tag{A.73}$$

where

$$\psi_{S_i} = \tan^{-1}(m_{S_i O}) \tag{A.74}$$

Table A.2 Coefficients of turning angle $\theta_{i3}^{L_{i3}}$ ($i = 1$ to 6)

Gait	Coefficients	Leg1	Leg2	Leg3	Leg4	Leg5	Leg6
DF = 1/2	m_2	3	0	0	3	3	0
	m_1	3	3	3	3	3	3
	K_1	1.5	1.5	1.5	1.5	1.5	1.5
DF = 2/3	m_2	4	1	2	-1	0	3
	m_1	2	2	2	2	2	2
	K_1	2	1	0	0	2	1
DF = 3/4	m_2	4.5	1.5	3	0	1.5	4.5
	m_1	1.5	1.5	1.5	1.5	1.5	1.5
	K_1	2.25	0.75	1.5	2.25	0.75	2.25

m_1 = no. of divisions of a gait cycle corresponding to swing angle,

m_2 = no. of divisions of a gait cycle covered before swing starts.

Table A.2 gives the numeric values of the different coefficients of turning angle $\theta_{i3}^{L_{i3}}$ ($i = 1$ to 6) at the start of leg swing during nth cycle for different gaits

II. *Crab Motion*

During crab motion, the swing leg is assumed to plan its trajectory in the XZ plane of local frame L_{i3}'' (refer to Fig. 3.4b).

(a) **Calculation of coordinates of points P_{i3}, Q_{i3}, T_{i3}, and P_{i3}' with respect to frame L_{i3}''**

Coordinates of P_{i3}

$$x_{P_{i3}}^{L_{i3}''} = 0; \tag{A.75}$$

$$y_{P_{i3}}^{L_{i3}''} = 0; \tag{A.76}$$

$$z_{P_{i3}}^{L_{i3}''} = 0 \tag{A.77}$$

Coordinates of Q_{i3}

$$z_{Q_{i3}}^{L_{i3}''} = H_{m_{i1}} + \Delta h \tag{A.78}$$

$$\Delta z_{P_{i3}Q_{i3}} = z_{Q_{i3}}^{L_{i3}''} - z_{P_{i3}}^{L_{i3}''} = H_{m_{i1}} + \Delta h \tag{A.79}$$

$$x_{Q_{i3}}^{L_{i3}''} = x_{P_{i3}}^{L_{i3}''} + \Delta z_{P_{i3}Q_{i3}} \cdot \tan \gamma_{xz} \tag{A.80}$$

Here $\gamma_{xz}|_a$ (given) is the angle of ascend of the trajectory in XZ plane.

Coordinates of T_{i3}

$$z_{T_{i3}}^{L_{i3}''} = z_{Q_{i3}}^{L_{i3}''} = \mathrm{H}_{m_{i1}} + \Delta h \tag{A.81}$$

$$\Delta z_{P_{i3}'T_{i3}} = z_{T_{i3}}^{L_{i3}''} - z_{P_{i3}'}^{L_{i3}''} = \mathrm{H}_{m_{i1}} + \Delta h - h_{i3}' \tag{A.82}$$

$$y_{T_{i3}}^{L_{i3}''} = 0 \tag{A.83}$$

$$x_{T_{i3}}^{L_{i3}''} = x_{P_{i3}'}^{L_{i3}''} - \Delta z_{P_{i3}'T_{i3}} . \tan \gamma_{xz}|_d \tag{A.84}$$

Here, $\gamma_{xz}|_d$ (given) is the angle of descend of the trajectory in XZ plane.

Coordinates of Point P'_{i3}

$$x_{P_{i3}'}^{L_{i3}''} = s_w (= s_0) \tag{A.85}$$

$$y_{P_{i3}'}^{L_{i3}''} = 0 \tag{A.86}$$

$$z_{P_{i3}'}^{L_{i3}''} = h_{i3}' \tag{A.87}$$

(b) **Swing leg trajectory planning with respect to frame L_{i3}''**

Along Z-axis:

$$z_{ei} = z_{ei}|_{t=t_0^s} + a_{z_i^{PQ}} . \Delta_{PQ}^2 . (3 - 2\Delta_{PQ}) \quad \text{for } t_{stort}^s \le t \le t_1^s \quad (\mathbf{P}_{i3} \text{ to } \mathbf{Q}_{i3}) \tag{A.88}$$

$$= z_{ei}|_{t=t_1^s} \quad \text{for } t_1^s \le t \le t_2^s \quad (\mathbf{Q}_{i3} \text{ to } \mathbf{T}_{i3}) \tag{A.89}$$

$$= z_{ei}|_{t=t_3^s} - a_{z_i^{TP'}} . \Delta_{TP'}^2 . (3 - 2\Delta_{TP'}) \quad \text{for } t_2^s \le t \le t_{end}^s \quad (\mathbf{T}_{i3} \text{ to } \mathbf{P}_{i3}') \tag{A.90}$$

where

$$z_{ei}|_{t=t_{start}^s} = z_{P_{i3}}^{L_{i3}''}, \ z_{ei}|_{t=t_1^s} = z_{Q_{i3}}^{L_{i3}''}, \ z_{ei}|_{t=t_2^s} = z_{T_{i3}}^{L_{i3}''}, \ z_{ei}|_{t=t_{end}^s} = z_{P_{i3}'}^{L_{i3}''} \tag{A.91}$$

$$a_{z_i^{PQ}} = z_{ei}|_{t=t_1^s} - z_{ei}|_{t=t_{start}^s} = z_{Q_{i3}}^{L_{i3}''} - z_{P_{i3}}^{L_{i3}''} \tag{A.92}$$

$$a_{z_i^{TP'}} = \left| z_{ei}|_{t=t_{end}^s} - z_{ei}|_{t=t_2^s} \right| = z_{T_{i3}}^{L_{i3}''} - z_{P_{i3}'}^{L_{i3}''} \tag{A.93}$$

$$\Delta_{PQ} = (t - t_{start}^s)/(t_1^s - t_{start}^s), \ \Delta_{TP'} = (t - t_2^s)/(t_{end}^s - t_2^s) \tag{A.94}$$

Along X-axis:

$$x_{ei} = x_{ei}|_{t=t_{start}^s} + a_{x_i^{PP'}}.\Delta_{PP'}^2.(3 - 2\Delta_{PP'}) \quad \text{for } t_{start}^s \le t \le t_{end}^s (\mathbf{P}_{i3} \text{ to } \mathbf{P}'_{i3}) \tag{A.95}$$

$$\dot{x}_{ei} = 6a_{x_i PP'}.\Delta_{PP'}(1 - \Delta_{PP'}).d\Delta_{PP'}/dt \tag{A.96}$$

$$\ddot{x}_{ei} = 6a_{x_i PP'}(1 - 2\Delta_{PP'})(d\Delta_{PP'}/dt)^2 \tag{A.97}$$

where

$$x_{ei}|_{t=t_{start}^s} = x_{P_{i3}}^{L_{i3}''}, \; x_{ei}|_{t=t_{end}^s} = x_{P_{i3}'}^{L_{i3}''} \tag{A.98}$$

$$a_{x_i PP'} = x_{ei}|_{t=t_{end}^s} - x_{ei}|_{t=t_{start}^s} = x_{P_{i3}'}^{L_{i3}''} - x_{P_{i3}}^{L_{i3}''} \tag{A.99}$$

$$\Delta_{PP'} = \left(t - t_{start}^s\right)/\left(t_{end}^s - t_{start}^s\right) \tag{A.100}$$

$$d\Delta_{PP'}/dt = 1/(t_{end}^s - t_{start}^s) \tag{A.101}$$

Superscript 's' indicates swing. For every swing phase of leg i, t_{start}^s, t_1^s, t_2^s and t_{end}^s indicate the start (point $\mathbf{P}_{i3}$), end of acceleration (point $\mathbf{Q}_{i3}$), start of deceleration (point $\mathbf{T}_{i3}$), and end (point $\mathbf{P}'_{i3}$) time, respectively.

Therefore, the coordinates of vector $\mathbf{r}_{P_{i3}^s P_{i3}}^{L_{i3}''}$ (refer to Eq. (3.82)) are given by,

$$\mathbf{r}_{P_{i3}^s P_{i3}}^{L_{i3}''} = (x_{ei}, \; 0, \; z_{ei})^T \tag{A.102}$$

Appendix A.5 Time Calculations for Gait Planning

I. Calculation of Total Time Taken to Complete n-duty Cycles

For all the n-duty cycles, the body stroke (s_0 or s_0^c) of the trunk body is assumed to be constant.

$$s_0 \text{ or } s_0^c = ms_0'' \tag{A.103}$$

Total displacement for n-cycles,

$$s_T = n.s_0 = n.s_0^c = n.m.s_0'' \tag{A.104}$$

Again,

Displacement during 1st cycle,

$$s_1 = (v_{t_0} + v_{t_0 t_1}/2).(1/\Delta'_a) + v_{t_1}(t_3^s\big|_1 - t_1) \tag{A.105}$$

(refer to Fig. 3.8).

where v_{t_0}, $v_{t_0 t_1}$ and v_{t_1} have their usual meanings as discussed in Sect. 3.2.6. It is to be noted that the displacement value during the velocity ramp is obtained by integrating Eq. (3.56) for straight-forward or Eq. (3.64) for crab or Eq. (3.51) for turning motion.
Likewise,

$$s_2 = v_{t_1}(t_3^s\big|_2 - t_3^s\big|_1) \tag{A.106}$$

$$\vdots \quad \vdots \quad \vdots$$

$$s_{n-1} = v_{t_1}(t_3^s\big|_{n-1} - t_3^s\big|_{n-2}) \tag{A.107}$$

$$s_n = (v_{t_1} - v_{t_0 t_1}/2).(1/\Delta'_d) + v_{t_1}(t_2 - t_3^s\big|_{n-1}) \tag{A.108}$$

Therefore, net displacement,

$$s_T = s_1 + s_2 + \cdots + s_{n-1} + s_n \tag{A.109}$$

$$= (v_{t_0} + v_{t_0 t_1}/2).(1/\Delta'_a) + (v_{t_1} - v_{t_0 t_1}/2).(1/\Delta'_d) + v_{t_1}(t_2 - t_1) \tag{A.110}$$

Comparing (A.104) and (A.110),

$$n.m.s_0'' = (v_{t_0} + v_{t_0 t_1}/2).(1/\Delta'_a) + (v_{t_1} - v_{t_0 t_1}/2).(1/\Delta'_d) + v_{t_1}(t_2 - t_1) \tag{A.111}$$

Substituting the values of t_1 and t_2 from Eq. (3.94) we get,

$$n.m.s_0'' = (v_{t_0} + v_{t_0 t_1}/2).(1/\Delta'_a) + (v_{t_1} - v_{t_0 t_1}/2).(1/\Delta'_d) + v_{t_1}((t_3 - \Delta t) - (t_0 + \Delta t)) \tag{A.112}$$

or $t_3 = t_0 + 2\Delta t + [n.m.s_0'' - (v_{t_0} + v_{t_0 t_1}/2).(1/\Delta'_a) - (v_{t_1} - v_{t_0 t_1}/2).(1/\Delta'_d)]/v_{t_1}$

II. *Calculation of End Time for Each of the Duty Cycles*

It is to be noted that displacement of trunk body during any cycle (linear or angular) is same, such that,

$$s_1 = s_2 = s_3 \cdots = s_n = m.s_0'' \tag{A.113}$$

Therefore,
1st cycle (time range t_0 to $t_3^s\big|_1$),

Comparing (A.113) and (A.105)

$$m.s_0'' = (v_{t_0} + v_{t_0 t_1}/2).(1/\Delta_a') + v_{t_1}(t_3^s\big|_1 - t_1)$$

$$\text{or } t_3^s\big|_1 = t_1 + [m.s_0'' - (v_{t_0} + v_{t_0 t_1}/2).(1/\Delta_a')]/v_{t_1} \qquad (A.114)$$

Similarly,
2nd cycle (time range $t_3^s\big|_1$ to $t_3^s\big|_2$),

$$t_3^s\big|_2 = t_3^s\big|_1 + m.s_0''/v_{t_1}$$

$$\vdots \qquad \vdots \qquad \vdots \qquad\qquad\qquad\qquad (A.115)$$

$n-1th$ cycle (time range $t_3^s\big|_{n-2}$ to $t_3^s\big|_{n-1}$),

$$t_3^s\big|_{n-1} = t_3^s\big|_{n-2} + m.s_0''/v_{t_1} \qquad (A.116)$$

nth cycle (time range $t_3^s\big|_{n-1}$ to $t_3^s\big|_n$),

$$t_3^s\big|_n = t_3^s\big|_{n-1} + [m.s_0'' - (v_{t_0} - v_{t_0 t_1}/2).(1/\Delta_d')]/v_{t_1} \qquad (A.117)$$

Appendix A.6 Kinematic Velocity and Acceleration

From equation set (3.23)

$$\mathbf{r}_{P_{11}O}^G = \mathbf{r}_{P_0 O}^G + \mathbf{A}^{GL_0}\mathbf{r}_{S_i P_0}^{L_0} + \mathbf{A}^{GL_i'}\mathbf{r}_{P_i S_i'}^{L_i'} \qquad (A.118)$$

Differentiating with respect to time,

$$\dot{\mathbf{r}}_{P_{i1}O}^G = \dot{\mathbf{r}}_{P_0 O}^G + \mathbf{A}^{GL_0}.{}^G\tilde{\boldsymbol{\omega}}_0.\mathbf{r}_{S_i P_0}^{L_0} + \mathbf{A}^{GL_i''}.{}^G\tilde{\boldsymbol{\omega}}_{i1}.\mathbf{r}_{P_{i1}S_i}^{L_i''}$$

$$= \dot{\mathbf{r}}_{P_0 O}^G - \mathbf{A}^{GL_0}.\tilde{\mathbf{r}}_{S_i P_0}^{L_0}.{}^G\boldsymbol{\omega}_0 - \mathbf{A}^{GL_i''}.\tilde{\mathbf{r}}_{P_{i1}S_i}^{L_i''}.{}^G\boldsymbol{\omega}_{i1} \qquad (A.119)$$

Calculation of Angular Velocities

For Trunk Body
The derivative of transformation matrix from frame L_0 to G is given by (refer to (Hahn 2003)),

$$\dot{\mathbf{A}}^{GL_0} = \left(\frac{\partial \mathbf{A}^{GL_0}}{\partial \alpha_0}.\dot{\alpha}_0 + \frac{\partial \mathbf{A}^{GL_0}}{\partial \beta_0}.\dot{\beta}_0 + \frac{\partial \mathbf{A}^{GL_0}}{\partial \theta_0} \cdot \dot{\theta}_0 \right)$$

$$c = \left(\frac{\partial \mathbf{A}^{GL_0}}{\partial \alpha_0}, \frac{\partial \mathbf{A}^{GL_0}}{\partial \beta_0}, \frac{\partial \mathbf{A}^{GL_0}}{\partial \theta_0} \right). \begin{pmatrix} \mathbf{I}_3.\dot{\alpha}_0 \\ \mathbf{I}_3.\dot{\beta}_0 \\ \mathbf{I}_3.\dot{\theta}_0 \end{pmatrix}$$

$$= \frac{\partial \mathbf{A}^{GL_0}}{\partial \boldsymbol{\eta}_0}.\dot{\boldsymbol{\eta}}_0 \qquad\qquad (A.120)$$

Now,

$$^{G}\tilde{\boldsymbol{\omega}}_0 = \mathbf{A}^{L_0 G}.\dot{\mathbf{A}}^{GL_0} \qquad\qquad (A.121)$$

After substitutions and arranging, we get angular velocity of trunk body as,

$$^{G}\boldsymbol{\omega}_0 = \begin{pmatrix} ^{G}\omega_{x0}, {}^{G}\omega_{y0}, {}^{G}\omega_{z0} \end{pmatrix}^T = \begin{pmatrix} \cos\beta_0 \cos\theta_0\dot{\alpha}_0 + \sin\theta_0\dot{\beta}_0 \\ -\cos\beta_0 \sin\theta_0\dot{\alpha}_0 + \cos\theta_0\dot{\beta}_0 \\ \sin\beta_0\dot{\alpha}_0 + \dot{\theta}_0 \end{pmatrix} = \mathbf{K}_0(\boldsymbol{\eta}_0).\dot{\boldsymbol{\eta}}_0,$$

$$\qquad\qquad (A.122)$$

where $\mathbf{K}_0(\boldsymbol{\eta}_0) \in \mathbb{R}^{3\times3}$ is the kinematic matrix associated with trunk body and

$$\dot{\boldsymbol{\eta}}_0 = \begin{pmatrix} \dot{\alpha}_0, \dot{\beta}_0, \dot{\theta}_0 \end{pmatrix}^T = \left(\frac{d\alpha_0}{dt}, \frac{d\beta_0}{dt}, \frac{d\theta_0}{dt} \right)^T = \frac{\partial \boldsymbol{\eta}_0}{\partial \mathbf{p}_0^G} \cdot \dot{\mathbf{p}}_0^G \qquad (A.123)$$

$$\mathbf{p}_0^G = ((\mathbf{r}_{P_0O}^G)^T, \boldsymbol{\eta}_0^T))^T = (x_{P_0O}^G \; y_{P_0O}^G \; z_{P_0O}^G \; \alpha_0 \; \beta_0 \; \theta_0)^T \quad \in \mathbb{R}^6 \qquad (A.124)$$

$$\dot{\mathbf{p}}_0^G = ((\dot{\mathbf{r}}_{P_0O}^G)^T, \dot{\boldsymbol{\eta}}_0^T))^T = (\dot{x}_{P_0O}^G \; \dot{y}_{P_0O}^G \; \dot{z}_{P_0O}^G \; \dot{\alpha}_0 \; \dot{\beta}_0 \; \dot{\theta}_0)^T \quad \in \mathbb{R}^6 \qquad (A.125)$$

For Link $i1$,
Similar to Eq. (A.120),

$$\dot{\mathbf{A}}^{GL_i''} = \frac{\partial \mathbf{A}^{GL_i''}}{\partial \boldsymbol{\eta}_{i1}}.\dot{\boldsymbol{\eta}}_{i1} \qquad\qquad (A.126)$$

and

$$^{G}\tilde{\boldsymbol{\omega}}_{i1} = \mathbf{A}^{L_i''G}.\dot{\mathbf{A}}^{GL_i''} \qquad\qquad (A.127)$$

Differentiating Eq. (3.6) with respect to time (substituting $j = 1$),

$$\dot{\boldsymbol{\eta}}_{i1} = \begin{pmatrix} \dot{\alpha}_{i1}, \dot{\beta}_{i1}, \dot{\theta}_{i1} \end{pmatrix}^T = \left(\frac{d\alpha_{i1}}{dt}, \frac{d\beta_{i1}}{dt}, \frac{d\theta_{i1}}{dt} \right)^T = \frac{\partial \boldsymbol{\eta}_{i1}}{\partial \mathbf{p}_0^G} \cdot \dot{\mathbf{p}}_0^G \qquad (A.128)$$

Therefore, angular velocity of link $i1$ is given by,

$$^G\boldsymbol{\omega}_{i1} = \left(^G\omega_{xi1}, {}^G\omega_{yi1}, {}^G\omega_{zi1}\right)^T = \mathbf{K}_{i1}(\boldsymbol{\eta}_{i1}).\dot{\boldsymbol{\eta}}_{i1} = \mathbf{K}_{i1}(\boldsymbol{\eta}_{i1}).\frac{\partial\boldsymbol{\eta}_{i1}}{\partial\mathbf{p}_0^G}.\dot{\mathbf{p}}_0^G = \mathbf{J}_{r_{i1}}.\dot{\mathbf{p}}_0^G$$

(A.129)

where

$$\mathbf{J}_{r_{i1}} = \mathbf{K}_{i1}(\boldsymbol{\eta}_{i1}).\frac{\partial\boldsymbol{\eta}_{i1}}{\partial\mathbf{p}_0^G} \quad \in \mathbb{R}^{3\times6}$$

(A.130)

$$= \begin{bmatrix} 0 & 0 & 0 & 0 & 0 & 0 \\ 0 & 0 & 0 & 0 & 0 & 0 \\ \frac{\partial\theta_{i1}}{\partial x_{P_0O}^G} & \frac{\partial\theta_{i1}}{\partial y_{P_0O}^G} & \frac{\partial\theta_{i1}}{\partial z_{P_0O}^G} & \frac{\partial\theta_{i1}}{\partial\alpha_0} & \frac{\partial\theta_{i1}}{\partial\beta_0} & \frac{\partial\theta_{i1}}{\partial\theta_0} \end{bmatrix}$$

(A.131)

$\mathbf{K}_{i1}(\boldsymbol{\eta}_{i1}) \in \mathbb{R}^{3\times3}$ is the kinematic matrix associated with link $i1$; $\frac{\partial\boldsymbol{\eta}_{i1}}{\partial\mathbf{p}_0^G} \in \mathbb{R}^{3\times6}$
Similarly, for the links $i2$ and $i3$,

$$^G\boldsymbol{\omega}_{i2} = \left(^G\omega_{xi2}, {}^G\omega_{yi2}, {}^G\omega_{zi2}\right)^T = \mathbf{J}_{r_{i2}}.\dot{\mathbf{p}}_0^G$$

(A.132)

$$^G\boldsymbol{\omega}_{i3} = \left(^G\omega_{xi3}, {}^G\omega_{yi3}, {}^G\omega_{zi3}\right)^T = \mathbf{J}_{r_{i3}}.\dot{\mathbf{p}}_0^G$$

(A.133)

Appendix A.7 Jacobian Matrices

$$\mathbf{J} = \begin{bmatrix} 0 & 0 & 0 & 0 & 0 & 0 \\ b_{21} & b_{22} & b_{23} & b_{24} & b_{25} & b_{26} \\ c_{31} & c_{32} & c_{33} & c_{34} & c_{35} & c_{36} \end{bmatrix}$$

(A.134)

For $i = 1$ to 6

1. For Jacobians $\mathbf{J} \equiv \mathbf{J}_{r_{ij}}$ and $\mathbf{J} \equiv \mathbf{J}'_{r_{ij}}$

    i. $j = 1$, $m = 1$ to 6,
        $b_{2m} = 0$,
        $c_{3m} = \partial\theta_{ij}/\partial V$,
        where $V \in \mathbf{p}_0^G$ for $\mathbf{J}_{r_{ij}}$ or $V \in \mathbf{p}_{i3}^G$ for $\mathbf{J}'_{r_{ij}}$
    ii. $j = 2$ to 3, $m = 1$ to 6,
        $b_{2m} = \partial\beta_{ij}/\partial V$,
        $c_{3m} = 0$,
        where $V \in \mathbf{p}_0^G$ for $\mathbf{J}_{r_{ij}}$ or $V \in \mathbf{p}_{i3}^G$ for $\mathbf{J}'_{r_{ij}}$

2. For Jacobians $\mathbf{J} \equiv \dot{\mathbf{J}}_{r_{ij}}$ and $\mathbf{J} \equiv \dot{\mathbf{J}}'_{r_{ij}}$

   i. $j = 1, m = 1$ to 6,

$$b_{2m} = 0,$$

$$c_{3m} = \dot{\mathbf{p}}_0^T . \mathbf{D}_{\theta_{ij}, V}, \text{ where } V \in \mathbf{p}_0^G \text{ for } \dot{\mathbf{J}}_{r_{ij}}$$

$$= \dot{\mathbf{p}}_0^T . \mathbf{D}_{\theta_{ij}, V} + \dot{\mathbf{p}}_{i3}^T . \mathbf{D}'_{\theta_{ij}, V}, \text{ where } V \in \mathbf{p}_{i3}^G \text{ for } \dot{\mathbf{J}}'_{r_{ij}}$$

   ii. $j = 2$ to 3, $m = 1$ to 6,

$$b_{2m} = \dot{\mathbf{p}}_0^T . \mathbf{D}_{\beta_{ij}, V}, \text{ where } V \in \mathbf{p}_0^G$$

$$= \dot{\mathbf{p}}_0^T . \mathbf{D}_{\beta_{ij}, V} + \dot{\mathbf{p}}_{i3}^T . \mathbf{D}'_{\beta_{ij}, V}, \text{ where } V \in \mathbf{p}_{i3}^G \text{ for } \dot{\mathbf{J}}'_{r_{ij}}$$

$$c_{3m} = 0,$$

Also,

$$\mathbf{D}_{A,V} = \left[\frac{\partial}{\partial x_{P_0 O}^G} \left(\frac{\partial A}{\partial V} \right) \; \frac{\partial}{\partial y_{P_0 O}^G} \left(\frac{\partial A}{\partial V} \right) \; \frac{\partial}{\partial z_{P_0 O}^G} \left(\frac{\partial A}{\partial V} \right) \; \frac{\partial}{\partial \alpha_0} \left(\frac{\partial A}{\partial V} \right) \; \frac{\partial}{\partial \beta_0} \left(\frac{\partial A}{\partial V} \right) \; \frac{\partial}{\partial \theta_0} \left(\frac{\partial A}{\partial V} \right) \right]^T$$

$$\mathbf{D}'_{A,C} = \left[\frac{\partial}{\partial x_{P_{i3} O}^G} \left(\frac{\partial A}{\partial V} \right) \; \frac{\partial}{\partial y_{P_{i3} O}^G} \left(\frac{\partial A}{\partial V} \right) \; \frac{\partial}{\partial z_{P_{i3} O}^G} \left(\frac{\partial A}{\partial V} \right) \; \frac{\partial}{\partial \alpha_{i3}} \left(\frac{\partial A}{\partial V} \right) \; \frac{\partial}{\partial \beta_{i3}} \left(\frac{\partial A}{\partial V} \right) \; \frac{\partial}{\partial \theta_{i3}} \left(\frac{\partial A}{\partial V} \right) \right]^T$$

$$A \in (\theta_{i1}, \beta_{i2}, \beta_{i3}),$$

$$\beta_{i3} = f\left(\mathbf{r}_{P_0 O}^G, \mathbf{r}_{P_{i3} O}^G, \boldsymbol{\eta}_0 \right)$$

$$\dot{\beta}_{i3} = \left[\frac{\partial \beta_{i3}}{\partial x_{P_0 O}^G} \; \frac{\partial \beta_{i3}}{\partial y_{P_0 O}^G} \; \frac{\partial \beta_{i3}}{\partial z_{P_0 O}^G} \; \frac{\partial \beta_{i3}}{\partial \alpha_0} \; \frac{\partial \beta_{i3}}{\partial \beta_0} \; \frac{\partial \beta_{i3}}{\partial \theta_0} \right] \left[\dot{x}_{P_0 O}^G \; \dot{y}_{P_0 O}^G \; \dot{z}_{P_0 O}^G \; \dot{\alpha}_0 \; \dot{\beta}_0 \; \dot{\theta}_0 \right]^T$$

$$+ \left[\frac{\partial \beta_{i3}}{\partial x_{P_{i3} O}^G} \; \frac{\partial \beta_{i3}}{\partial y_{P_{i3} O}^G} \; \frac{\partial \beta_{i3}}{\partial z_{P_{i3} O}^G} \right] \left[\dot{x}_{P_{i3} O}^G \; \dot{y}_{P_{i3} O}^G \; \dot{z}_{P_{i3} O}^G \right]^T$$

$$\ddot{\beta}_{i3} = \left[\frac{\partial \beta_{i3}}{\partial x_{P_0 O}^G} \; \frac{\partial \beta_{i3}}{\partial y_{P_0 O}^G} \; \frac{\partial \beta_{i3}}{\partial z_{P_0 O}^G} \; \frac{\partial \beta_{i3}}{\partial \alpha_0} \; \frac{\partial \beta_{i3}}{\partial \beta_0} \; \frac{\partial \beta_{i3}}{\partial \theta_0} \right] \left[\ddot{x}_{P_0 O}^G \; \ddot{y}_{P_0 O}^G \; \ddot{z}_{P_0 O}^G \; \ddot{\alpha}_0 \; \ddot{\beta}_0 \; \ddot{\theta}_0 \right]^T$$

$$+ \left[\frac{\partial \beta_{i3}}{\partial x_{P_{i3} O}^G} \; \frac{\partial \beta_{i3}}{\partial y_{P_{i3} O}^G} \; \frac{\partial \beta_{i3}}{\partial z_{P_{i3} O}^G} \right] \left[\ddot{x}_{P_{i3} O}^G \; \ddot{y}_{P_{i3} O}^G \; \ddot{z}_{P_{i3} O}^G \right]^T$$

$$+ \left[\frac{\partial^2 \beta_{i3}}{\partial \left(x_{P_0 O}^G \right)^2} \; \frac{\partial^2 \beta_{i3}}{\partial \left(y_{P_0 O}^G \right)^2} \; \frac{\partial^2 \beta_{i3}}{\partial \left(z_{P_0 O}^G \right)^2} \; \frac{\partial^2 \beta_{i3}}{\partial \alpha_0^2} \; \frac{\partial^2 \beta_{i3}}{\partial \beta_0^2} \; \frac{\partial^2 \beta_{i3}}{\partial \theta_0^2} \right]$$

$$\times \left[\left(\dot{x}_{P_0 O}^G \right)^2 \; \left(\dot{y}_{P_0 O}^G \right)^2 \; \left(\dot{z}_{P_0 O}^G \right)^2 \; \dot{\alpha}_0^2 \; \dot{\beta}_0^2 \; \dot{\theta}_0^2 \right]^T$$

$$+ \left[\frac{\partial^2 \beta_{i3}}{\partial \left(x_{P_{i3} O}^G \right)^2} \; \frac{\partial^2 \beta_{i3}}{\partial \left(y_{P_{i3} O}^G \right)^2} \; \frac{\partial^2 \beta_{i3}}{\partial \left(z_{P_{i3} O}^G \right)^2} \right] . \left[\left(\dot{x}_{P_{i3} O}^G \right)^2 \; \left(\dot{y}_{P_{i3} O}^G \right)^2 \; \left(\dot{z}_{P_{i3} O}^G \right)^2 \right]^T$$

Appendix A.8 Parameters Affecting the Dynamics of the Six-Legged Robot

Summary of the parameters affecting the dynamics of the six-legged robot taken with respect to G_0 [refer to (Hahn 2002)].

1. *Mass matrix of the rigid body 'k':*

$$\mathbf{M}_k(\mathbf{p}_k^{G_0}) = \begin{bmatrix} m_k.\mathbf{I}_3 & -m_k.\mathbf{A}^{G_0L}.\tilde{\mathbf{r}}_{C_k P_k}^L \\ m_k.\left(\tilde{\mathbf{r}}_{C_k P_k}^L\right)^T.\mathbf{A}^{LG_0} & \mathbf{J}_{P_k}^L \end{bmatrix}, \in \mathbb{R}^{6,6} \qquad (A.135)$$

where moment of inertia,

$$\mathbf{J}_{P_k}^L = \mathbf{J}_{C_k}^L + m_k.\left\{ \left(\mathbf{r}_{C_k P_k}^L\right)^T.\mathbf{r}_{C_k P_k}^L.\mathbf{I}_3 - \mathbf{r}_{C_k P_k}^L.\left(\mathbf{r}_{C_k P_k}^L\right)^T \right\} \qquad (A.136)$$

such that,

$$\mathbf{J}_{C_k}^L = \begin{bmatrix} J_{C_k x}^L & -J_{C_k xy}^L & -J_{C_k xz}^L \\ -J_{C_k yx}^L & J_{C_k y}^L & -J_{C_k yz}^L \\ -J_{C_k zx}^L & -J_{C_k zy}^L & J_{C_k z}^L \end{bmatrix} \qquad (A.137)$$

and

$$\tilde{\mathbf{r}}_{C_k P_k}^L = \begin{bmatrix} 0 & -z_{C_k P_k} & y_{C_k P_k} \\ z_{C_k P_k} & 0 & -x_{C_k P_k} \\ -y_{C_k P_k} & x_{C_k P_k} & 0 \end{bmatrix} \qquad (A.138)$$

$\mathbf{I}_3$ is the 3 by 3 identity matrix.

2. *Acceleration matrix of the rigid body 'k':*

$$\dot{\mathbf{v}}_k^{G_0} = \begin{bmatrix} {}^{G_0}\ddot{\mathbf{r}}_{P_k O}^{G_0} \\ {}^{G_0}\boldsymbol{\omega}_k \end{bmatrix}, \in \mathbb{R}^6 \qquad (A.139)$$

3. *Applied forces and moments of the rigid body 'k':*

Applied forces,

$$\mathbf{F}_{W_k}^{G_0} = \begin{bmatrix} 0 & 0 & -m_k g \end{bmatrix}^T \qquad (A.140)$$

Applied moments,

Table A.3 Unknown forces and moments acting in different components of the six-legged robot

Components	F_{kx}	F_{ky}	F_{kz}	M_{kx}	M_{ky}	M_{kz}
Trunk body and payload	0	0	0	0	0	$M_{11} + M_{21} + M_{31} + M_{41} + M_{51} + M_{61}$
Link $i1$ ($j = 1$)	0	0	0	0	M_{i2}	$-M_{i1}$
Link $i2$ ($j = 2$)	0	0	0	0	$-M_{i2} + M_{i3}$	0
Link $i3$ ($j = 3$)	$F_{ix}^{G_0}$	$F_{iy}^{G_0}$	$F_{iz}^{G_0}$	$T_{ix}^{G_0}$	$-M_{i3} + T_{iy}^{G_0}$	$T_{iz}^{G_0}$

NB: $F_{ix}^{G_0} = {}^m F_{ix}^{G_0} + F_{ix}^{\prime G_0}$; $F_{iy}^{G_0} = {}^m F_{iy}^{G_0} + F_{iy}^{\prime G_0}$; $F_{iz}^{G_0} = {}^m F_{iz}^{G_0} + F_{iz}^{\prime G_0}$

$$\mathbf{M}_{W_k}^L = \tilde{\mathbf{r}}_{C_k P_k}^L . \mathbf{F}_{W_k}^L, \in \mathbb{R}^3 \tag{A.141}$$

where

$$\mathbf{F}_{W_k}^L = \mathbf{A}^{LG_0} . \mathbf{F}_{W_k}^{G_0}, \in \mathbb{R}^3 \tag{A.142}$$

Total applied forces and moment,

$$\mathbf{f}_k \left(\mathbf{p}_k^{G_0}, \mathbf{v}_k^{G_0} \right) = \mathbf{f}_{K_k} + \mathbf{f}_{U_k} \tag{A.143}$$

where
Known forces and moments,

$$\mathbf{f}_{K_k} = \begin{bmatrix} 0 & 0 & -m_k g & M_{W_k x}^L & M_{W_k y}^L & M_{W_k z}^L \end{bmatrix}^T \tag{A.144}$$

Unknown forces and moments,

$$\mathbf{f}_{U_k} = [\, F_{kx}\, F_{ky}\, F_{kz}\, M_{kx}\, M_{ky}\, M_{kz}\,]^T \tag{A.145}$$

(refer to Table A.3)

4. *Centrifugal forces and gyroscopic terms of the rigid body 'k':*

$$\mathbf{q}_{G_0 C_k} = \begin{bmatrix} -m_k . \mathbf{A}^{G_0 L} . {}^{G_0}\tilde{\boldsymbol{\omega}}_k . {}^{G_0}\tilde{\boldsymbol{\omega}}_k . \mathbf{r}_{C_k P_k}^L \\ -{}^{G_0}\tilde{\boldsymbol{\omega}}_k . \mathbf{J}_{P_k}^L . {}^{G_0}\boldsymbol{\omega}_k \end{bmatrix} \in \mathbb{R}^6 \tag{A.146}$$

Corollary 1

(i) $L = L_0$ for $k = 0$, designates trunk body and payload (combined)

Combined COM of trunk body and payload,

$$\mathbf{r}_{C_0 P_0}^{L_0} = \frac{m_B \mathbf{r}_{C_B P_0}^{L_0} + m_L \mathbf{r}_{C_L P_0}^{L_0}}{m_0} \quad \in \mathbb{R}^3 \tag{A.147}$$

Combined mass of trunk body and payload,

$$m_0 = m_B + m_L \tag{A.148}$$

where $\mathbf{C}_T$ and $\mathbf{C}_L$ are the locations of the COM of the trunk body and payload, respectively; $\mathbf{r}_{C_B P_0}^{L_0}$ and $\mathbf{r}_{C_L P_0}^{L_0}$ are the positions of $\mathbf{C}_B$ and $\mathbf{C}_L$ from $\mathbf{P}_0$ with respect to frame L_0; m_B and m_L are the masses of trunk and payload, respectively.

Mass moment of inertia:

$$J_{C_0 x}^{L_0} = J_{C_B x}^{L_0} + m_B \left(x_{C_0 P_0}^{L_0} - x_{C_B P_0}^{L_0} \right)^2 + J_{C_L x}^{L_0} + m_L \left(x_{C_0 P_0}^{L_0} - x_{C_L P_0}^{L_0} \right)^2 \tag{A.149}$$

$$J_{C_0 y}^{L_0} = J_{C_B y}^{L_0} + m_B \left(y_{C_0 P_0}^{L_0} - y_{C_B P_0}^{L_0} \right)^2 + J_{C_L y}^{L_0} + m_L \left(y_{C_0 P_0}^{L_0} - y_{C_L P_0}^{L_0} \right)^2 \tag{A.150}$$

$$J_{C_0 z}^{L_0} = J_{C_B z}^{L_0} + m_B \left(z_{C_0 P_0}^{L_0} - z_{C_B P_0}^{L_0} \right)^2 + J_{C_L z}^{L_0} + m_L \left(z_{C_0 P_0}^{L_0} - z_{C_L P_0}^{L_0} \right)^2 \tag{A.151}$$

$$J_{C_0 yx}^{L_0} = J_{C_0 xy}^{L_0} = J_{C_B xy}^{L_0} + m_B \left(x_{C_0 P_0}^{L_0} - x_{C_B P_0}^{L_0} \right) \left(y_{C_0 P_0}^{L_0} - y_{C_B P_0}^{L_0} \right)$$
$$+ J_{C_L xy}^{L_0} + m_L \left(x_{C_0 P_0}^{L_0} - x_{C_L P_0}^{L_0} \right) \left(y_{C_0 P_0}^{L_0} - y_{C_L P_0}^{L_0} \right) \tag{A.152}$$

$$J_{C_0 zy}^{L_0} = J_{C_0 yz}^{L_0} = J_{C_B yz}^{L_0} + m_B \left(y_{C_0 P_0}^{L_0} - y_{C_B P_0}^{L_0} \right) \left(z_{C_0 P_0}^{L_0} - z_{C_B P_0}^{L_0} \right)$$
$$+ J_{C_L yz}^{L_0} + m_L \left(y_{C_0 P_0}^{L_0} - y_{C_L P_0}^{L_0} \right) \left(z_{C_0 P_0}^{L_0} - z_{C_L P_0}^{L_0} \right) \tag{A.153}$$

$$J_{C_0 xz}^{L_0} = J_{C_0 zx}^{L_0} = J_{C_B zx}^{L_0} + m_B \left(z_{C_0 P_0}^{L_0} - z_{C_B P_0}^{L_0} \right) \left(x_{C_0 P_0}^{L_0} - x_{C_B P_0}^{L_0} \right)$$
$$+ J_{C_L zx}^{L_0} + m_L \left(z_{C_0 P_0}^{L_0} - z_{C_L P_0}^{L_0} \right) \left(x_{C_0 P_0}^{L_0} - x_{C_L P_0}^{L_0} \right) \tag{A.154}$$

(ii) $\mathbf{L} = \mathbf{L}'_{ij}$ for $k = ij$ designates link ij, for $i = 1$ to $6, j = 1$ to 3. Here, C_{ij} is the location of the COM of link ij; m_{ij} is the mass of link ij.

Total summation of applied forces and moments:

$$\mathbf{f}\left(\mathbf{p}^{G_0}, \mathbf{v}^{G_0}\right) = \mathbf{f}_0 \left(\mathbf{p}_0^{G_0}, \mathbf{v}_0^{G_0}\right) + \sum_i^6 \sum_j^3 \mathbf{f}_{ij}\left(\mathbf{p}_{ij}^{G_0}, \mathbf{v}_{ij}^{G_0}\right) \text{ for } i = 1 \text{ to } 6, \ j = 1 \text{ to } 3$$

$$= \mathbf{f}_{K_0} + \mathbf{f}_{U_0} + \sum_i^6 \sum_j^3 \left(\mathbf{f}_{K_{ij}} + \mathbf{f}_{U_{ij}}\right)$$

$$= \mathbf{f}_{K_0} + \sum_i^6 \sum_j^3 \mathbf{f}_{K_{ij}} + \mathbf{f}_{U_0} + \sum_i^6 \sum_j^3 \mathbf{f}_{U_{ij}}$$

$$= \mathbf{f}_K + \mathbf{f}_U \tag{A.155}$$

where
Summation of known forces and moments,

$$\mathbf{f}_K = \mathbf{f}_{K_0} + \sum_i^6 \sum_j^3 \mathbf{f}_{K_{ij}} \tag{A.156}$$

Summation of unknown forces and moments,

$$\mathbf{f}_U = \mathbf{f}_{U_0} + \sum_i^6 \sum_j^3 \mathbf{f}_{U_{ij}} \tag{A.157}$$

Total summation of centrifugal and gyroscopic terms:

$$\mathbf{q}_{GC}(\mathbf{p}^{G_0}, \mathbf{v}^{G_0}) = \mathbf{q}_{G_0 C_0} + \sum_i^6 \sum_j^3 \mathbf{q}_{G_0 C_{ij}} \tag{A.158}$$

Appendix A.9 Kinematic constraints with respect to G_0

1. *Position Constraints with respect to* G_0

For *ith* leg,

$$\begin{pmatrix} \mathbf{p}_0^{G_0} \\ \mathbf{p}_i^{G_0} \end{pmatrix}^{24 \times 1} = \begin{pmatrix} \mathbf{r}_{P_0 O}^{G_0} \\ \boldsymbol{\eta}_0 \\ \mathbf{r}_{P_0 O}^{G_0} + \mathbf{A}^{G_0 L_0} \mathbf{r}_{S_i P_0}^{L_0} + \mathbf{A}^{G_0 L_i''} \mathbf{r}_{P_{i1} S_i}^{L_i''} \\ \mathbf{P}_r(z).\mathbf{P}_r^T(z).\boldsymbol{\eta}_{i1} \\ \mathbf{r}_{P_0 O}^{G_0} + \mathbf{A}^{G_0 L_0} \mathbf{r}_{S_i P_0}^{L_0} + \mathbf{A}^{G_0 L_i''} \mathbf{r}_{P_{i1} S_i}^{L_i''} + \mathbf{A}^{G_0 L_{i1}''} \mathbf{r}_{P_{i2} P_{i1}}^{L_{i1}''} \\ \mathbf{P}_r(y).\mathbf{P}_r^T(y).\boldsymbol{\eta}_{i2} \\ \mathbf{r}_{P_0 O}^{G_0} + \mathbf{A}^{G_0 L_0} \mathbf{r}_{S_i P_0}^{L_0} + \mathbf{A}^{G_0 L_i''} \mathbf{r}_{P_{i1} S_i}^{L_i''} + \mathbf{A}^{G_0 L_{i1}''} \mathbf{r}_{P_{i2} P_{i1}}^{L_{i1}''} + \mathbf{A}^{G_0 L_{i2}''} \mathbf{r}_{P_{i3} P_{i2}}^{L_{i2}''} \\ \mathbf{P}_r(y).\mathbf{P}_r^T(y).\boldsymbol{\eta}_{i3} \end{pmatrix}^{24 \times 1} \tag{A.159}$$

In Eq. (A.159), $\mathbf{P}_r$ denotes the matrix projectors; $\mathbf{r}_{S_i P_0}^{L_0}, \mathbf{r}_{P_{i1} S_i}^{L_i''}, \mathbf{r}_{P_{i2} P_{i1}}^{L_{i1}''}, \mathbf{r}_{P_{i3} P_{i2}}^{L_{i2}''}$ are the local position of points S_i, $\mathbf{P}_{i1}$, $\mathbf{P}_{i2}$ and $\mathbf{P}_{i3}$ with respect to frames L_0, L''_i, L''_{i1} and L''_{i2} respectively and are fixed for a robotic structure; $\mathbf{A}^{G_0 L_0}, \mathbf{A}^{GL_i'}, \mathbf{A}^{GL_i''}, \mathbf{A}^{GL_{i1}''},$

$\mathbf{A}^{GL''_{i2}}$, $\mathbf{A}^{GL''_{i2}}$ are the transformation matrices related to different local frames with respect to frame $\mathbf{G}$. $\mathbf{A}^{L''_i G}$, $\mathbf{A}^{L''_{i1} G}$, $\mathbf{A}^{L''_{i2} G}$ are orthogonal matrices of $\mathbf{A}^{GL''_i}$, $\mathbf{A}^{GL''_{i1}}$, $\mathbf{A}^{GL''_{i2}}$, respectively.

2. Velocity Constraints with respect to $\mathbf{G}_0$

For ith leg,

$$\left(\begin{matrix} \mathbf{v}_0^{G_0} \\ \mathbf{v}_i^{G_0} \end{matrix} \right)^{24\times 1} = \left(\begin{array}{c} \dot{\mathbf{r}}_{P_0 O}^{G_0} \\ {}^{G_0}\boldsymbol{\omega}_0 \\ \dot{\mathbf{r}}_{P_0 O}^{G_0} - \mathbf{A}^{G_0 L_0} \tilde{\mathbf{r}}_{S_i P_0}^{L_0} \cdot {}^{G_0}\boldsymbol{\omega}_0 - \mathbf{A}^{G_0 L''_i} \tilde{\mathbf{r}}_{P_{i1} S_i}^{L''_i} \cdot {}^{G_0}\boldsymbol{\omega}_{i1} \\ \mathbf{P}_r(z).\mathbf{P}_r^T(z).{}^{G_0}\boldsymbol{\omega}_{i1} \\ \dot{\mathbf{r}}_{P_0 O}^{G_0} - \mathbf{A}^{GL_0} \tilde{\mathbf{r}}_{S_i P_0}^{L_0} \cdot {}^{G_0}\boldsymbol{\omega}_0 - \mathbf{A}^{G_0 L''_i} \tilde{\mathbf{r}}_{P_{i1} S_i}^{L''_i} \cdot {}^{G_0}\boldsymbol{\omega}_{i1} - \mathbf{A}^{G_0 L''_{i1}} \tilde{\mathbf{r}}_{P_{i2} P_{i1}}^{L''_{i1}} \cdot {}^{G_0}\boldsymbol{\omega}_{i2} \\ \mathbf{P}_r(y).\mathbf{P}_r^T(y).{}^{G_0}\boldsymbol{\omega}_{i2} \\ \dot{\mathbf{r}}_{P_0 O}^{G_0} - \mathbf{A}^{G_0 L_0} \tilde{\mathbf{r}}_{S_i P_0}^{L_0} \cdot {}^{G_0}\boldsymbol{\omega}_0 - \mathbf{A}^{G_0 L''_i} \tilde{\mathbf{r}}_{P_{i1} S_i}^{L''_i} \cdot {}^{G_0}\boldsymbol{\omega}_{i1} - \mathbf{A}^{G_0 L''_{i1}} \tilde{\mathbf{r}}_{P_{i2} P_{i1}}^{L''_{i1}} \cdot {}^{G_0}\boldsymbol{\omega}_{i2} - \mathbf{A}^{G_0 L''_{i2}} \tilde{\mathbf{r}}_{P_{i3} P_{i2}}^{L''_{i2}} \cdot {}^{G_0}\boldsymbol{\omega}_{i3} \\ \mathbf{P}_r(y).\mathbf{P}_r^T(y).{}^{G_0}\boldsymbol{\omega}_{i3} \end{array} \right)^{24\times 1}$$

(A.160)

$$\left(\begin{matrix} \mathbf{v}_0^{G_0} \\ \mathbf{v}_i^{G_0} \end{matrix} \right)^{24\times 1} = \left[\begin{array}{ccccc} \mathbf{I}_3 & 0_{3,3} & 0_{3,3} & 0_{3,3} & 0_{3,3} \\ 0_{3,3} & \mathbf{I}_3 & 0_{3,3} & 0_{3,3} & 0_{3,3} \\ \mathbf{I}_3 & -\mathbf{A}^{G_0 L_0}.\tilde{\mathbf{r}}_{S_i P_0}^{L_0} & -\mathbf{A}^{G_0 L''_i}.\tilde{\mathbf{r}}_{P_{i1} S_i}^{L''_i} & 0_{3,3} & 0_{3,3} \\ 0_{3,3} & 0_{3,3} & \mathbf{P}_r(z).\mathbf{P}_r^T(z) & 0_{3,3} & 0_{3,3} \\ \mathbf{I}_3 & -\mathbf{A}^{G_0 L_0}.\tilde{\mathbf{r}}_{S_i P_0}^{L_0} & -\mathbf{A}^{G_0 L''_i}.\tilde{\mathbf{r}}_{P_{i1} S_i}^{L''_i} & -\mathbf{A}^{G_0 L''_{i1}}.\tilde{\mathbf{r}}_{P_{i2} P_{i1}}^{L''_{i1}} & 0_{3,3} \\ 0_{2,3} & 0_{2,3} & 0_{2,3} & \mathbf{P}_r(y).\mathbf{P}_r^T(y) & 0_{3,3} \\ \mathbf{I}_3 & 0_{3,3} & 0_{3,3} & 0_{3,3} & -\mathbf{A}^{G_0 L''_{i2}}.\tilde{\mathbf{r}}_{P_{i3} P_{i2}}^{L''_{i2}} \\ 0_{2,3} & 0_{2,3} & 0_{2,3} & 0_{2,3} & \mathbf{P}_r(y).\mathbf{P}_r^T(y) \end{array} \right]^{24\times 15} \cdot \left(\begin{matrix} \dot{\mathbf{r}}_{P_0 O}^{G_0} \\ {}^{G_0}\boldsymbol{\omega}_0 \\ {}^{G_0}\boldsymbol{\omega}_{i1} \\ {}^{G_0}\boldsymbol{\omega}_{i2} \\ {}^{G_0}\boldsymbol{\omega}_{i3} \end{matrix} \right)^{15\times 1}$$

Also,

(A.161)

$$^{G_0}\boldsymbol{\omega}_{i1} = [0\ 0\ {}^{G_0}\omega_{i1z}]^T$$
$$^{G_0}\boldsymbol{\omega}_{i2} = [0\ {}^{G_0}\omega_{i2y}\ 0]^T$$
$$^{G_0}\boldsymbol{\omega}_{i3} = [0\ {}^{G_0}\omega_{i3y}\ 0]^T$$

(A.162)

$$\mathbf{P}_r(y) = [0\ 1\ 0]^T, \mathbf{P}_r(z) = [0\ 0\ 1]^T$$

(A.163)

Hence,

$$\mathbf{A}^{G_0 L''_i} \tilde{\mathbf{r}}_{P_{i1} S_i}^{L''_i} \cdot {}^{G_0}\boldsymbol{\omega}_{i1} = \mathbf{t}_{i1}.{}^{G_0}\omega_{i1z}$$

(A.164)

$$\mathbf{A}^{G_0 L''_{i1}} \cdot \tilde{\mathbf{r}}_{P_{i2} P_{i1}}^{L''_{i1}} \cdot {}^{G_0}\boldsymbol{\omega}_{i2} = \mathbf{t}_{i2}.{}^{G_0}\omega_{i2y}$$

(A.165)

$$\mathbf{A}^{G_0 L''_{i2}} \cdot \tilde{\mathbf{r}}_{P''_{i3} P_{i2}}^{L''_{i2}} \cdot {}^{G_0}\boldsymbol{\omega}_{i3} = \mathbf{t}_{i3} \cdot {}^{G_0}\omega_{i3y}$$

(A.166)

where $\mathbf{t}_{i1}, \mathbf{t}_{i2}, \mathbf{t}_{i3} \in \mathbb{R}^{3 \times 1}$

And

$$\mathbf{P}_r(z) . \mathbf{P}_r^T(z) .^{G_0} \boldsymbol{\omega}_{i1} = \mathbf{s}_{iz} .^{G_0} \omega_{i1z} \tag{A.167}$$

$$\mathbf{P}_r(y) . \mathbf{P}_r^T(y) .^{G_0} \boldsymbol{\omega}_{i2} = \mathbf{s}_{iy} .^{G_0} \omega_{i2y} \tag{A.168}$$

$$\mathbf{P}_r(y) . \mathbf{P}_r^T(y) .^{G_0} \boldsymbol{\omega}_{i3} = \mathbf{s}_{iy} .^{G_0} \omega_{i3y} \tag{A.169}$$

where

$$\mathbf{s}_{iy} = [0\ 1\ 0]^T, \mathbf{s}_{iz} = [0\ 0\ 1]^T, \tag{A.170}$$

Substituting Eqs. (A.163)–(A.170) in Eq. (A.161),

$$\begin{pmatrix} \mathbf{v}_0^{G_0} \\ \mathbf{v}_i^{G_0} \end{pmatrix}^{24 \times 1} = \begin{bmatrix} \mathbf{I}_6 & \mathbf{0}_{6,3} \\ \mathbf{R}_{0i} & \mathbf{R}_i \end{bmatrix}^{24 \times 9} . \begin{pmatrix} \mathbf{u}_0 \\ \mathbf{u}_i \end{pmatrix}^{9 \times 1} \tag{A.171}$$

where

$$\mathbf{v}_i^{G_0} \left(\dot{\mathbf{p}}_0^{G_0}, \dot{\mathbf{p}}_i^{G_0} \right) = \mathbf{J}_i(\mathbf{q}_0, \mathbf{q}_i) \cdot \mathbf{u}_i(\dot{\mathbf{q}}_0, \dot{\mathbf{q}}_i) \tag{A.172}$$

where $\mathbf{q}_0$ and $\mathbf{q}_i$ are the vectors of generalized coordinates.

$$\mathbf{u}_0 = ((\dot{\mathbf{r}}_{P_0 O}^{G_0})^T, (^{G_0} \boldsymbol{\omega}_0)^T)^T \in \mathbb{R}^6 \tag{A.173}$$

$$\mathbf{u}_i = [\,^{G_0} \omega_{i1z} \quad ^{G_0} \omega_{i2y} \quad ^{G_0} \omega_{i3y}\,]^T \in \mathbb{R}^3 \tag{A.174}$$

$$\mathbf{R}_{0i} = \begin{bmatrix} \mathbf{I}_3 & -\mathbf{A}^{G_0 L_0} . \tilde{\mathbf{r}}_{S_i P_0}^{L_0} \\ \mathbf{0}_{3,3} & \mathbf{0}_{3,3} \\ \mathbf{I}_3 & -\mathbf{A}^{G_0 L_0} . \tilde{\mathbf{r}}_{S_i P_0}^{L_0} \\ \mathbf{0}_{3,3} & \mathbf{0}_{3,3} \\ \mathbf{I}_3 & -\mathbf{A}^{G_0 L_0} . \tilde{\mathbf{r}}_{S_i P_0}^{L_0} \\ \mathbf{0}_{3,3} & \mathbf{0}_{3,3} \end{bmatrix}^{18 \times 6} \quad ; \tag{A.175}$$

$$\mathbf{R}_i = \begin{bmatrix} -\mathbf{t}_{i1} & \mathbf{0}_{3,1} & \mathbf{0}_{3,1} \\ \mathbf{s}_{iz} & \mathbf{0}_{3,1} & \mathbf{0}_{3,1} \\ -\mathbf{t}_{i1} & -\mathbf{t}_{i2} & \mathbf{0}_{3,1} \\ \mathbf{0}_{3,1} & \mathbf{s}_{iy} & \mathbf{0}_{3,1} \\ -\mathbf{t}_{i1} & -\mathbf{t}_{i2} & -\mathbf{t}_{i3} \\ \mathbf{0}_{3,1} & \mathbf{0}_{3,1} & \mathbf{s}_{iy} \end{bmatrix}^{18 \times 3} \tag{A.176}$$

Overall velocity constraint set of the system with respect to G_0 is defined by,

$$
\begin{pmatrix}
\mathbf{v}_0^{G_0} \\
\mathbf{v}_1^{G_0} \\
\mathbf{v}_2^{G_0} \\
\mathbf{v}_3^{G_0} \\
\mathbf{v}_4^{G_0} \\
\mathbf{v}_5^{G_0} \\
\mathbf{v}_6^{G_0}
\end{pmatrix}^{114\times 1}
=
\begin{bmatrix}
\mathbf{I}_6 & \mathbf{0}_{6,3} & \mathbf{0}_{6,3} & \mathbf{0}_{6,3} & \mathbf{0}_{6,3} & \mathbf{0}_{6,3} & \mathbf{0}_{6,3} \\
\mathbf{R}_{01} & \mathbf{R}_1 & \mathbf{0}_{18,3} & \mathbf{0}_{18,3} & \mathbf{0}_{18,3} & \mathbf{0}_{18,3} & \mathbf{0}_{18,3} \\
\mathbf{R}_{02} & \mathbf{0}_{18,3} & \mathbf{R}_2 & \mathbf{0}_{18,3} & \mathbf{0}_{18,3} & \mathbf{0}_{18,3} & \mathbf{0}_{18,3} \\
\mathbf{R}_{03} & \mathbf{0}_{18,3} & \mathbf{0}_{18,3} & \mathbf{R}_3 & \mathbf{0}_{18,3} & \mathbf{0}_{18,3} & \mathbf{0}_{18,3} \\
\mathbf{R}_{04} & \mathbf{0}_{18,3} & \mathbf{0}_{18,3} & \mathbf{0}_{18,3} & \mathbf{R}_4 & \mathbf{0}_{18,3} & \mathbf{0}_{18,3} \\
\mathbf{R}_{05} & \mathbf{0}_{18,3} & \mathbf{0}_{18,3} & \mathbf{0}_{18,3} & \mathbf{0}_{18,3} & \mathbf{R}_5 & \mathbf{0}_{18,3} \\
\mathbf{R}_{06} & \mathbf{0}_{18,3} & \mathbf{0}_{18,3} & \mathbf{0}_{18,3} & \mathbf{0}_{18,3} & \mathbf{0}_{18,3} & \mathbf{R}_6
\end{bmatrix}^{114\times 24}
\cdot
\begin{pmatrix}
\mathbf{u}_0 \\
\mathbf{u}_1 \\
\mathbf{u}_2 \\
\mathbf{u}_3 \\
\mathbf{u}_4 \\
\mathbf{u}_5 \\
\mathbf{u}_6
\end{pmatrix}^{24\times 1}
$$
$$
=: \mathbf{v}^{G_0} \qquad\qquad =: \mathbf{J} \qquad\qquad =: \mathbf{u}
$$
$$(A.177)$$

3. *Acceleration constraints with respect to G_0*

Differentiating Eq. (A.177) with respect to time gives,

$$
\begin{pmatrix}
\dot{\mathbf{v}}_0^{G_0} \\
\dot{\mathbf{v}}_i^{G_0}
\end{pmatrix}^{24\times 1}
=
\begin{bmatrix}
\mathbf{I}_6 & \mathbf{0}_{6,3} \\
\mathbf{R}_{0i} & \mathbf{R}_i
\end{bmatrix}^{24\times 9}
\cdot
\begin{pmatrix}
\dot{\mathbf{u}}_0 \\
\dot{\mathbf{u}}_i
\end{pmatrix}^{9\times 1}
+
\begin{pmatrix}
\mathbf{0}_6 \\
\mathbf{N}_i
\end{pmatrix}^{24\times 1}
\qquad (A.178)
$$

or

$$
\dot{\mathbf{v}}_i^{G_0}\left(\ddot{\mathbf{p}}_0^{G_0}, \ddot{\mathbf{p}}_i^{G_0}\right) = \mathbf{J}_i(\mathbf{q}_0, \mathbf{q}_i) \cdot \dot{\mathbf{u}}_i(\ddot{\mathbf{q}}_0, \ddot{\mathbf{q}}_i) + \dot{\mathbf{J}}_i(\dot{\mathbf{q}}_0, \dot{\mathbf{q}}_i) \cdot \mathbf{u}_i(\dot{\mathbf{q}}_0, \dot{\mathbf{q}}_i) \qquad (A.179)
$$

$$
\dot{\mathbf{u}}_0 = ((\ddot{\mathbf{r}}_{P_0O}^{G_0})^T, (^{G_0}\dot{\boldsymbol{\omega}}_0)^T)^T \in \mathbb{R}^6 \qquad (A.180)
$$

$$
\dot{\mathbf{u}}_i = [\,^{G_0}\dot{\omega}_{i1z} \;\; ^{G_0}\dot{\omega}_{i2y} \;\; ^{G_0}\dot{\omega}_{i3y}\,]^T \in \mathbb{R}^3 \qquad (A.181)
$$

$$
\mathbf{N}_i =
\begin{pmatrix}
\mathbf{A}^{G_0L_0}\cdot {}^{G_0}\tilde{\boldsymbol{\omega}}_0 \cdot {}^{G_0}\tilde{\boldsymbol{\omega}}_0 \cdot \mathbf{r}_{S_iP_0}^{L_0} + \mathbf{A}^{G_0L_i^*}\cdot {}^{G_0}\tilde{\boldsymbol{\omega}}_{i1}\cdot {}^{G_0}\tilde{\boldsymbol{\omega}}_{i1}\cdot \mathbf{r}_{P_1S_i}^{L_i^*} \\
\mathbf{0}_{3,1} \\
\mathbf{A}^{G_0L_0}\cdot {}^{G_0}\tilde{\boldsymbol{\omega}}_0 \cdot {}^{G_0}\tilde{\boldsymbol{\omega}}_0 \cdot \mathbf{r}_{S_iP_0}^{L_0} + \mathbf{A}^{G_0L_i^*}\cdot {}^{G_0}\tilde{\boldsymbol{\omega}}_{i1}\cdot {}^{G_0}\tilde{\boldsymbol{\omega}}_{i1}\cdot \mathbf{r}_{P_1S_i}^{L_i^*} + \mathbf{A}^{G_0L_{i1}}\cdot {}^{G_0}\tilde{\boldsymbol{\omega}}_{i2}\cdot {}^{G_0}\tilde{\boldsymbol{\omega}}_{i2}\cdot \mathbf{r}_{P_2P_1}^{L_{i1}} \\
\mathbf{0}_{3,1} \\
\mathbf{A}^{G_0L_0}\cdot {}^{G_0}\tilde{\boldsymbol{\omega}}_0 \cdot {}^{G_0}\tilde{\boldsymbol{\omega}}_0 \cdot \mathbf{r}_{S_iP_0}^{L_0} + \mathbf{A}^{G_0L_i^*}\cdot {}^{G_0}\tilde{\boldsymbol{\omega}}_{i1}\cdot {}^{G_0}\tilde{\boldsymbol{\omega}}_{i1}\cdot \mathbf{r}_{P_1S_i}^{L_i^*} + \mathbf{A}^{G_0L_{i1}}\cdot {}^{G_0}\tilde{\boldsymbol{\omega}}_{i2}\cdot {}^{G_0}\tilde{\boldsymbol{\omega}}_{i2}\cdot \mathbf{r}_{P_2P_1}^{L_{i1}} + \mathbf{A}^{G_0L_{i2}}\cdot {}^{G_0}\tilde{\boldsymbol{\omega}}_{i3}\cdot {}^{G_0}\tilde{\boldsymbol{\omega}}_{i3}\cdot \mathbf{r}_{P_3P_2}^{L_{i2}} \\
\mathbf{0}_{3,1}
\end{pmatrix}^{18\times 1}
$$
$$(A.182)$$

Overall acceleration constraint set of the system with respect to G_0 is defined by,

$$
\overset{114\times1}{\begin{pmatrix} \dot{\mathbf{v}}^{G_0}_0 \\ \dot{\mathbf{v}}^{G_0}_1 \\ \dot{\mathbf{v}}^{G_0}_2 \\ \dot{\mathbf{v}}^{G_0}_3 \\ \dot{\mathbf{v}}^{G_0}_4 \\ \dot{\mathbf{v}}^{G_0}_5 \\ \dot{\mathbf{v}}^{G_0}_6 \end{pmatrix}} = \overset{114\times24}{\begin{bmatrix} \mathbf{I}_6 & \mathbf{0}_{6,3} & \mathbf{0}_{6,3} & \mathbf{0}_{6,3} & \mathbf{0}_{6,3} & \mathbf{0}_{6,3} & \mathbf{0}_{6,3} \\ \mathbf{R}_{01} & \mathbf{R}_1 & \mathbf{0}_{18,3} & \mathbf{0}_{18,3} & \mathbf{0}_{18,3} & \mathbf{0}_{18,3} & \mathbf{0}_{18,3} \\ \mathbf{R}_{02} & \mathbf{0}_{18,3} & \mathbf{R}_2 & \mathbf{0}_{18,3} & \mathbf{0}_{18,3} & \mathbf{0}_{18,3} & \mathbf{0}_{18,3} \\ \mathbf{R}_{03} & \mathbf{0}_{18,3} & \mathbf{0}_{18,3} & \mathbf{R}_3 & \mathbf{0}_{18,3} & \mathbf{0}_{18,3} & \mathbf{0}_{18,3} \\ \mathbf{R}_{04} & \mathbf{0}_{18,3} & \mathbf{0}_{18,3} & \mathbf{0}_{18,3} & \mathbf{R}_4 & \mathbf{0}_{18,3} & \mathbf{0}_{18,3} \\ \mathbf{R}_{05} & \mathbf{0}_{18,3} & \mathbf{0}_{18,3} & \mathbf{0}_{18,3} & \mathbf{0}_{18,3} & \mathbf{R}_5 & \mathbf{0}_{18,3} \\ \mathbf{R}_{06} & \mathbf{0}_{18,3} & \mathbf{0}_{18,3} & \mathbf{0}_{18,3} & \mathbf{0}_{18,3} & \mathbf{0}_{18,3} & \mathbf{R}_6 \end{bmatrix}} \cdot \overset{24\times1}{\begin{pmatrix} \dot{\mathbf{u}}_0 \\ \dot{\mathbf{u}}_1 \\ \dot{\mathbf{u}}_2 \\ \dot{\mathbf{u}}_3 \\ \dot{\mathbf{u}}_4 \\ \dot{\mathbf{u}}_5 \\ \dot{\mathbf{u}}_6 \end{pmatrix}} + \overset{24\times1}{\begin{pmatrix} \mathbf{0}_6 \\ \mathbf{N}_1 \\ \mathbf{N}_2 \\ \mathbf{N}_3 \\ \mathbf{N}_4 \\ \mathbf{N}_5 \\ \mathbf{N}_6 \end{pmatrix}}
$$

$$
=: \dot{\mathbf{v}}^{G_0} \qquad\qquad =: \mathbf{J} \qquad\qquad\qquad =: \dot{\mathbf{u}} \qquad =: \dot{\mathbf{J}} \cdot \mathbf{u}
$$

$$(A.183)$$

Appendix A.10 Geometrical Interpretation of the Interaction Region

Considering $\triangle ABH$ (refer to Fig. A.3),

$$\angle BAH = \delta_i$$

$$HA = a = p/\sin\delta_i \qquad (A.184)$$

$$HB = b = p/\cos\delta_i \qquad (A.185)$$

Here, p is the maximum depth of penetration, and δ_i is the angle subtended by the bottom face of the footpad with the ground and is expressed by,

$$\delta_i = \pi/2 - (\beta_2 + \beta_{13} - \psi) \qquad (A.186)$$

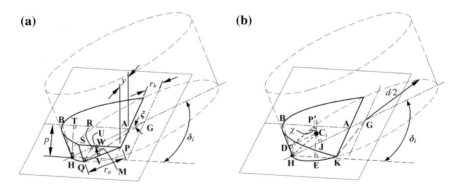

Fig. A.3 Feet–ground interaction between leg-tip (rubber pad) and ground **a** segmented volume, **b** total volume

where β_{i2} and β_{i3} are the angular displacement of the joints and ψ is the bounding angle expressed by (refer to Mahapatra et al. 2015),

$$\psi = \tan^{-1}\left(d/2l'_{i3}\right) \tag{A.187}$$

Here, d is the diameter of footpad and l'_{i3} is the length of the tibia from rotation axis to the center of footpad, i.e., point G.

Also,

$$\sin \delta_i = p/(d/2 - r_b) \tag{A.188}$$

where r_b is the distance from the chord of interception to the center of circle of the bottom face of the footpad.

Thus,

$$AG = r_b = d/2 - p/\sin \delta_i \tag{A.189}$$

Let y be the perpendicular distance of the $\triangle PQS$ from the center of footpad as shown in Fig. A.3a.

From $\triangle GRQ$,

$$\sin \xi = QR/QG = 2y/d \tag{A.190}$$

and

$$\cos \xi = QR/QG = \sqrt{d^2 - 4y^2}/d \tag{A.191}$$

Thus,

$$PQ = r_a = QG.\cos \xi - AG = d/2.\cos \xi - r_b \tag{A.192}$$

Substituting Eq. (A.191) in Eq. (A.192),

$$PQ = r_a = \frac{1}{2}\sqrt{d^2 - 4y^2} - r_b \tag{A.193}$$

Since $\triangle PQS$ and $\triangle ABH$ are equivalent triangles,

$$\angle BHA = \angle SQP = \pi/2$$

and

$$\angle BAH = \angle SPQ = \delta_i$$

$$SQ = PQ.\tan\delta_i = r_a.\tan\delta_i = \left(\frac{1}{2}\sqrt{d^2 - 4y^2} - r_b\right).\tan\delta_i \qquad (A.194)$$

From $\triangle AGK$,

$$AK = \sqrt{(d/2)^2 - r_b^2} \qquad (A.195)$$

Area of the intercepted $\triangle PQS$ is given by,

$$A_y = 1/2.\ SQ.PQ = \frac{1}{2}PQ^2.\tan\delta_i = \frac{1}{2}r_a^2.\tan\delta_i \qquad (A.196)$$

Volume of a slice of infinitesimal thickness with this triangular cross-section PQS,

$$dV = A_y.dy$$

Then, volume of the interaction region,

$$V = \oint dV = \int_{-AK}^{+AK} A_y dy = \int_{-AK}^{+AK} \frac{1}{2}r_a^2.\tan\delta_i.dy = \tan\delta_i.\int_0^{\sqrt{(d/2)^2 - r_b^2}} r_a^2.dy \qquad (A.197)$$

Considering the segment $\triangle PQS$ in Fig. A.3a, whose centroid is at M and the depth into the ground being the length WM (designated by h),

$$UV = \frac{2}{3}SQ,\ VM = \frac{1}{3}SQ, \qquad (A.198)$$

And,

$$UM = UV - VM = \frac{1}{3}SQ \qquad (A.199)$$

$$\angle UMW = \delta_i \qquad (A.200)$$

Now,

$$WM = UM \cdot \cos\delta_i = \frac{1}{3}SQ \cdot \cos\delta_i \qquad (A.201)$$

Substituting Eq. (A.194) in Eq. (A.201),

$$WM = r_a \cdot \sin\delta_i/3 \qquad (A.202)$$

Therefore, the depth of COM of the contact volume is given by,

$$P'_{i3}C_i = h_c = \oint \mathrm{WM} \cdot dV / \oint dV \tag{A.203}$$

Substituting Eqs. (A.202) and (A.197) in (A.203),

$$h_c = \sin \delta_i . \int_0^{\sqrt{(d/2)^2 - r_b^2}} r_a^3 \cdot dy/3 . \int_0^{\sqrt{(d/2)^2 - r_b^2}} r_a^2 . dy \tag{A.204}$$

Substituting Eq. (A.193) in Eq. (A.204),

$$h_c = \frac{\sin \delta_i}{32} \left[\frac{3d^2 (16r_b^2 + d^2) \cos^{-1}(2r_b/d) - 2r_b(8r_b^2 + 13d^2)\sqrt{d^2 - 4r_b^2}}{(d^2 + 2r_b^2)\sqrt{d^2 - 4r_b^2} - 3r_b d^2 \cos^{-1}(2r_b/d)} \right] \tag{A.205}$$

Similarly, support distance of the segment $\triangle PQS$ is given by,

$$VP = 2/3r_a \tag{A.206}$$

For entire contact volume,

$$JA = r_e = \oint \mathrm{VP} \cdot dV / \oint dV \tag{A.207}$$

Substituting Eqs. (A.206) and (A.197) in Eq. (A.207),

$$r_e = 2 . \int_0^{\sqrt{(d/2)^2 - r_b^2}} r_a^3 \cdot dy/3 . \int_0^{\sqrt{(d/2)^2 - r_b^2}} r_a^2 . dy \tag{A.208}$$

Substituting Eq. (A.204) in Eq. (A.208),

$$r_e = 2h_c / \sin \delta_i \tag{A.209}$$

$$HJ = DC_i = d/2 - r_b - r_e \tag{A.210}$$

In $\triangle DC_i P'_{i3}$,
Let

$$\angle DP'_{i3}C_i = \chi \tag{A.211}$$

From trigonometrical relationships,

$$\angle DC_i P'_{i3} = \pi/2 + \delta_i \tag{A.212}$$

From laws of triangles,

$$P'_{i3}D = w = \sqrt{h_c^2 + (d/2 - r_b - r_e)^2 + 2h_c(d/2 - r_b - r_e)\sin\delta_i} \tag{A.213}$$

$$\cos\chi = \frac{P'_{i3}D^2 + P'_{i3}C_i^2 - DC_i^2}{2P'_{i3}D.P'_{i3}C_i} \tag{A.214}$$

$$P'_{i3}D^2 = P'_{i3}C_i^2 + DC_i^2 + 2P'_{i3}C_i \cdot DC_i \cdot \sin\delta_i \tag{A.215}$$

Substituting Eq. (A.215) in (A.214),

$$\chi = \cos^{-1}\big((P'_{i3}C_i + DC_i \cdot \sin\delta_i/P'_{i3}D)\big) = (h_c + (d/2 - r_b - r_e) \cdot \sin\delta_i)/w \tag{A.216}$$

Appling laws of triangles to $\Delta BD\,P'_{i3}$,

$$BD = P'_{i3}D \cdot \cos\chi/\cos\delta_i = w \cdot \cos\chi/\cos\delta_i \tag{A.217}$$

Now,

$$HD = HB - BD = (p - w \cdot \cos\chi)/\cos\delta_i \tag{A.218}$$

Appendix A.11 Objective Function and Evaluation of the Constraints

The appendix describes how the torque minimization problem is put into the form of a QP problem in the current work. The overall joint torque vector is represented by,

$$\mathbf{M}^{G_0} = \big[\,(\mathbf{M}_1^{G_0})^T (\mathbf{M}_2^{G_0})^T (\mathbf{M}_3^{G_0})^T (\mathbf{M}_4^{G_0})^T (\mathbf{M}_5^{G_0})^T (\mathbf{M}_6^{G_0})^T\,\big]^T \in \mathbb{R}^{18} \tag{A.219}$$

Further, $\mathbf{M}_i^{G_0}$ ($i = 1$ to 6) is the torque vector of each joint and is the function of primary variables, which in this case are the foot forces $F_{ix}^{G_0}$, $F_{iy}^{G_0}$, and $F_{iz}^{G_0}$. It is obtained by substituting Eqs. (A.73) and (A.62) in Eq. (A.37) such that,

$$\mathbf{M}_i^{G_0} = -\mathbf{B}_i^{-1}\big(\mathbf{A}_i\mathbf{F}_i^{G_0} + \mathbf{D}_i\mathbf{T}_i^{G_0} + \mathbf{M}_{ei}^{G_0}\big)$$
$$= \mathbf{U}_i\mathbf{F}_i^{G_0} + \mathbf{V}_i \tag{A.220}$$

where

$$\mathbf{T}_i^{G_0} = \mathbf{A}^{G_0 N_{i3}} \tilde{\mathbf{r}}_i^{N_{i3}} \mathbf{A}^{GG_0} \mathbf{F}_i^{G_0} \in \mathbb{R}^3 \qquad (A.221)$$

$$\mathbf{U}_i = -\mathbf{B}_i^{-1} \left(\mathbf{A}_i + \mathbf{D}_i \mathbf{A}^{G_0 N_{i3}} \tilde{\mathbf{r}}_i^{N_{i3}} \mathbf{A}^{GG_0} \right) \in \mathbb{R}^3 \qquad (A.222)$$

$$\mathbf{V}_i = -\mathbf{B}_i^{-1} \mathbf{M}_{ei}^{G_0} \in \mathbb{R}^3 \qquad (A.223)$$

Likewise, the overall joint torque vector,

$$\mathbf{M}^{G_0} = \mathbf{U}\mathbf{F}^{G_0} + \mathbf{V} \qquad (A.224)$$

where

$$\mathbf{U} = \begin{bmatrix} \mathbf{U}_1 & \mathbf{0}_3 & \dots & \mathbf{0}_3 \\ \mathbf{0}_3 & \mathbf{U}_2 & \dots & \mathbf{0}_3 \\ \vdots & \vdots & \ddots & \vdots \\ \mathbf{0}_3 & \mathbf{0}_3 & \dots & \mathbf{U}_6 \end{bmatrix} \in \mathbb{R}^{18 \times 18} \qquad (A.225)$$

$$\mathbf{F}^{G_0} = \left[(\mathbf{F}_1^{G_0})^T (\mathbf{F}_2^{G_0})^T (\mathbf{F}_3^{G_0})^T (\mathbf{F}_4^{G_0})^T (\mathbf{F}_5^{G_0})^T (\mathbf{F}_6^{G_0})^T \right]^T \in \mathbb{R}^{18} \qquad (A.226)$$

$$\mathbf{V} = \left[\mathbf{V}_1^T \mathbf{V}_2^T \mathbf{V}_3^T \mathbf{V}_4^T \mathbf{V}_5^T \mathbf{V}_6^T \right]^T \in \mathbb{R}^{18} \qquad (A.227)$$

Now, substituting Eq. (A.224) in Eq. (3.70), the objective function can be written as,

$$\begin{aligned} S(\mathbf{T}^{G_0}) &= \frac{1}{2} \left(\mathbf{U}\mathbf{F}^{G_0} + \mathbf{V} \right)^T \mathbf{W} \left(\mathbf{U}\mathbf{F}^{G_0} + \mathbf{V} \right) \\ &= \frac{1}{2} \left((\mathbf{F}^{G_0})^T \mathbf{U}^T \mathbf{W}\mathbf{U}\mathbf{F}^{G_0} + \mathbf{V}^T \mathbf{W}\mathbf{U}\mathbf{F}^{G_0} + (\mathbf{F}^{G_0})^T \mathbf{U}^T \mathbf{W}\mathbf{V} + \mathbf{V}^T \mathbf{W}\mathbf{V} \right) \\ &= \frac{1}{2} \left((\mathbf{F}^{G_0})^T \mathbf{U}^T \mathbf{W}\mathbf{U}\mathbf{F}^{G_0} + (\mathbf{U}^T \mathbf{W}\mathbf{V})^T \mathbf{F}^{G_0} + (\mathbf{U}^T \mathbf{W}\mathbf{V})^T \mathbf{F}^{G_0} + \mathbf{V}^T \mathbf{W}\mathbf{V} \right) \\ &= \frac{1}{2} \left((\mathbf{F}^{G_0})^T \mathbf{U}^T \mathbf{W}\mathbf{U}\mathbf{F}^{G_0} + (\mathbf{U}^T \mathbf{W}\mathbf{V})^T \mathbf{F}^{G_0} + (\mathbf{V}^T \mathbf{W}\mathbf{V}) \right) \\ &= \frac{1}{2} (\mathbf{F}^{G_0})^T \bar{\mathbf{H}}\mathbf{F}^{G_0} + \mathbf{p}^T \mathbf{F}^{G_0} + \frac{1}{2} \mathbf{V}^T \mathbf{W}\mathbf{V} \end{aligned} \qquad (A.228)$$

($\mathbf{W}$ is a symmetric matrix)
where

$$\bar{\mathbf{H}} = \mathbf{U}^T \mathbf{W}\mathbf{U}, \qquad (A.229)$$

$$\mathbf{c} = \mathbf{U}^T \mathbf{W}\mathbf{V} \qquad (A.230)$$

The last term of Eq. (A.228) does not depend on $\mathbf{F}^{G_0}$. So the new objective function can be described by a new equation,

$$S(\mathbf{F}^{G_0}) = \frac{1}{2}\left(\mathbf{F}^{G_0}\right)^T \bar{\mathbf{H}}\mathbf{F}^{G_0} + \mathbf{c}^T \mathbf{F}^{G_0} \tag{A.231}$$

1. *Force-moment equality constraints*:

Equation (4.68) in matrix form is written as,

$$\tilde{\mathbf{r}}_{C_m O}^{G_0}\mathbf{F}_{C_m}^{G_0} + \sum (\mathbf{T}_i^{G_0} + \tilde{\mathbf{s}}_i^{G_0}\mathbf{F}_i^{G_0}) = \mathbf{0}_3 \tag{A.232}$$

Substituting Eq. (A.221) in (A.232) and rearranging,

$$\sum (\mathbf{A}^{G_0 N_{i3}}\tilde{\mathbf{r}}_i^{N_{i3}}\mathbf{A}^{GG_0} + \tilde{\mathbf{s}}_i^{G_0})\mathbf{F}_i^{G_0} = -\tilde{\mathbf{r}}_{C_m O}^{G_0}\mathbf{F}_{C_m}^{G_0}$$

or

$$\sum \mathbf{K}_i\mathbf{F}_i^{G_0} = -\tilde{\mathbf{r}}_{C_m O}^{G_0}\mathbf{F}_{C_m}^{G_0} \tag{A.233}$$

where

$$\mathbf{K}_i = \mathbf{A}^{G_0 N_{i3}}\tilde{\mathbf{r}}_i^{N_{i3}}\mathbf{A}^{GG_0} + \tilde{\mathbf{s}}_i^{G_0} \in \mathbb{R}^3 \tag{A.234}$$

Therefore, for all the six legs,

$$\mathbf{K}\mathbf{F}^{G_0} = -\tilde{\mathbf{r}}_{C_m O}^{G_0}\mathbf{F}_{C_m^{G_m}}^{G_0} \tag{A.235}$$

where

$$\mathbf{K} = \begin{bmatrix} \tilde{\mathbf{K}}_1 & \tilde{\mathbf{K}}_2 & \tilde{\mathbf{K}}_3 & \tilde{\mathbf{K}}_4 & \tilde{\mathbf{K}}_5 & \tilde{\mathbf{K}}_6 \end{bmatrix} \in \mathbb{R}^{3\times 18} \tag{A.236}$$

Rearranging Eq. (4.17),

$$\mathbf{I}_0\mathbf{F}^{G_0} = -\mathbf{F}_e^{G_0} \tag{A.237}$$

where

$$\mathbf{I}_0 = \begin{bmatrix} \mathbf{I}_3 & \mathbf{I}_3 & \mathbf{I}_3 & \mathbf{I}_3 & \mathbf{I}_3 & \mathbf{I}_3 \end{bmatrix} \in \mathbb{R}^{3\times 18} \tag{A.238}$$

$\mathbf{I}_3$ is an identity matrix.
Similarly Eq. (4.18) reduces to,

$$\mathbf{R}\mathbf{F}^{G_0} = -\mathbf{M}_0^{G_0} - \mathbf{M}_e^{G_0} \tag{A.239}$$

where

$$\mathbf{R} = \begin{bmatrix} \tilde{\mathbf{s}}_1^{G_0} & \tilde{\mathbf{s}}_2^{G_0} & \tilde{\mathbf{s}}_3^{G_0} & \tilde{\mathbf{s}}_4^{G_0} & \tilde{\mathbf{s}}_5^{G_0} & \tilde{\mathbf{s}}_6^{G_0} \end{bmatrix} \in \mathbb{R}^{3 \times 18} \qquad (A.240)$$

Combining equality constraint Eqs. (A.235), (A.237), and (A.239), the overall set of equality constraints is given by:

$$\mathbf{A}_e.\mathbf{F}^{G_0} = \mathbf{B}_e \qquad (A.241)$$

where

$$\mathbf{A}_e = \begin{pmatrix} \mathbf{K} \\ \mathbf{I}_0 \\ \mathbf{R} \end{pmatrix}, \in \mathbb{R}^{9 \times 18} \qquad (A.242)$$

$$\mathbf{B}_e = -\begin{pmatrix} \tilde{\mathbf{r}}_{C_m}^{G_0} {}_O \mathbf{F}_{C_m}^{G_0} \\ \mathbf{F}_e^{G_0} \\ \mathbf{M}_0^{G_0} + \mathbf{M}_e^{G_0} \end{pmatrix} \in \mathbb{R}^9 \qquad (A.243)$$

2. *Joint torque inequality constraints*:

For each joint (Eq. (4.69)),

$$M_{ij,min} \leq M_{ij}^{G_0} \leq M_{ij,max} \qquad (A.244)$$

For each leg i,

$$\begin{pmatrix} M_{i1,min} \\ M_{i2,min} \\ M_{i3,min} \end{pmatrix} \leq \begin{pmatrix} M_{i1}^{G_0} \\ M_{i2}^{G_0} \\ M_{i3}^{G_0} \end{pmatrix} \leq \begin{pmatrix} M_{i1,max} \\ M_{i2,max} \\ M_{i3,max} \end{pmatrix} \qquad (A.245)$$

$$\mathbf{M}_{i,min} \leq \mathbf{M}_i^{G_0} \leq \mathbf{M}_{i,max} \in \mathbb{R}^3 \qquad (A.246)$$

For all the legs,

$$\mathbf{M}_{min} \leq \mathbf{M}^{G_0} \leq \mathbf{M}_{max} \in \mathbb{R}^{18} \qquad (A.247)$$

Substituting Eq. (A.224) in Eq. (A.247),

$$\mathbf{M}_{min} \leq \mathbf{U}\mathbf{F}^{G_0} + \mathbf{V} \leq \mathbf{M}_{max} \qquad (A.248)$$

which means,

$$\mathbf{U}\mathbf{F}^{G_0} \geq \mathbf{M}_{min} - \mathbf{V} \qquad (A.249)$$

and

$$-\mathbf{U}\mathbf{F}^{G_0} \geq \mathbf{V} - \mathbf{M}_{\max} \tag{A.250}$$

3. *Friction force inequality constraints*:

Combining Eqs. (4.55) and (4.60),

$$\mathbf{Q}_i \mathbf{A}^{GG_0} \mathbf{F}_i^{G_0} \geq \mathbf{0}_4 \tag{A.251}$$

Therefore, composite friction force inequality constraints for all the legs,

$$\mathbf{Q}\mathbf{F}^{G_0} \geq \mathbf{0}_{24} \tag{A.252}$$

where

$$\mathbf{Q} = \begin{bmatrix} \mathbf{Q}_1.\mathbf{A}^{GG_0} & \mathbf{0}_3 & \cdots & \mathbf{0}_3 \\ \mathbf{0}_3 & \mathbf{Q}_2.\mathbf{A}^{GG_0} & \cdots & \mathbf{0}_3 \\ \vdots & \vdots & \ddots & \vdots \\ \mathbf{0}_3 & \mathbf{0}_3 & \cdots & \mathbf{Q}_6.\mathbf{A}^{GG_0} \end{bmatrix} \in \mathbb{R}^{24 \times 18} \tag{A.253}$$

Combining Eqs. (A.249), (A.250), and (A.252), the overall set of inequality constraints is given by:

$$\mathbf{A}_u \cdot \mathbf{F}^{G_0} \geq \mathbf{B}_u \tag{A.254}$$

where

$$\mathbf{A}_u = \begin{pmatrix} \mathbf{U} \\ -\mathbf{U} \\ \mathbf{Q} \end{pmatrix} \in \mathbb{R}^{60 \times 18} \tag{A.255}$$

$$\mathbf{B}_u = \begin{pmatrix} \mathbf{M}_{\min} - \mathbf{V} \\ \mathbf{V} - \mathbf{M}_{\max} \\ \mathbf{0}_{24} \end{pmatrix} \in \mathbb{R}^{60} \tag{A.256}$$

Index

© Springer Nature Singapore Pte Ltd. 2020
A. Mahapatra et al., *Multi-body Dynamic Modeling of Multi-legged Robots*,
Cognitive Intelligence and Robotics, https://doi.org/10.1007/978-981-15-2953-5

Printed in the United States
By Bookmasters